Environmental Discourses in Science Education

Volume 6

This series wrestles with the tensions situated between environmental and science education and addresses the scholarly efforts to bring confluence to these two projects with the help of ecojustice philosophy. As ecojustice is one of the fastest emerging trends for evaluating science education policy, the topics addressed in this series can help guide pedagogical trends such as critical media literacy, citizen science, and activism. The series emphasizes ideological analysis, curriculum studies and research in science educational policy, where there is a need for recognizing the tensions between cultural and natural systems, the way language is endorsed within communities and associated influence, and morals and ethics embedded in school science. Conversations and new perspectives on residual issues within science education are likely to be addressed in nuanced ways when considering the significance of ecojustice, defensible environmentalism, free-choice. Book proposals for this series may be submitted to the Publishing Editor: Claudia Acuna E-mail: Claudia.Acuna@springer.com

More information about this series at http://www.springer.com/series/11800

Joel B. Pontius • Michael P. Mueller
David Greenwood

Editors

Place-based Learning for the Plate

Hunting, Foraging and Fishing for Food

Editors
Joel B. Pontius
Sustainability and Environmental
Education Department
Goshen College
Goshen, IN, USA

Michael P. Mueller
College of Education
University of Alaska Anchorage
Anchorage, AK, USA

David Greenwood
Lakehead University
Thunder Bay, ON, Canada

ISSN 2352-7307   ISSN 2352-7315   (electronic)
Environmental Discourses in Science Education
ISBN 978-3-030-42813-6   ISBN 978-3-030-42814-3   (eBook)
https://doi.org/10.1007/978-3-030-42814-3

This Springer imprint is published by the registered company Springer Nature Switzerland AG.
The registered company address is: Gewerbestrasse 11, 6330 Cham, Switzerland

Preface

…when the food does not come from a flock in the sky, when you don't feel the warm feathers cool in your hand and know that a life has been given for yours, when there is no gratitude in return – that food may not satisfy. It may leave the spirit hungry while the belly is full…. A great longing is upon us, to live again in a world made of gifts. (Robin Wall Kimmerer (2013, p. 30–31))

Gifts

When I was 9, I received an incredible gift along the thunderstorm-flooded shoreline of a small reservoir in North Central Indiana. I had ridden my red bike, with a lime green 5-gallon bucket hanging on the handlebars, to this place I knew. As I walked across the dam, I noticed that the water level was high and hoped they had come up for air.

And they had. One after another, I carefully grasped the crayfish with my thumb and index finger, just behind their raised claws, and dropped them into the bucket. I had eaten lobster before and guessed that these creatures would share a common flavor. There were 3 dozens crawling over each other in the bucket as I peddled home.

I helped my dad cook them. As we lowered them into boiling water, the crayfish transformed instantly from dark brown with black gleaming eyes to pinkish red over their whole bodies. I thought the color was a good sign, as I knew lobsters changed color when they were cooked. With a steaming crayfish on a saucer next to the stovetop, I pulled the tail, which separated easily from the body, and removed the exoskeleton like I had with shrimp at Grandpa Verch's house. I took a cautious bite and chewed for a second. They tasted like lobster, but sweeter.

There was enough food for my whole family. My brother and I shelled the tails, piling them on a serving plate, and my dad worked the meat out of the claws with an old nutcracker. While my older sister would usually pass on fish I had caught or rabbit stew from hunts along the railroad tracks with dad, she dipped the plump white tails into lemon butter and went back for more. This was a different kind of meal. I felt like I had been let in on something special, a miracle that would happen only once. This was one of the many childhood gathering experiences that encouraged me to continue paying close attention to the places around me and to approach them with curiosity and openness. Decades later, the learning continues to deepen, reveal, disrupt, and reorient.

Home

"They look like beautiful flowers, Papa!" my 6-year-old daughter, Emmy, yelled as she leapt into the air with an orange mushroom pinched between her fingers. When we walk slowly through the forest together along the ridge top path west of our house, it seems as though the land throws open its deep pockets. This past summer, we learned seven new varieties of edible mushrooms within a few hundred steps of our door, including cinnabar red chanterelles, the orange mushrooms we were gathering. These days, our homeplace in the headwaters of Lake Michigan is one of our most influential teachers. Through the long growing seasons, there is something to learn each day, and the discoveries – especially the ones that make their way to our table – blur the walls of our nylon-sided house with the colors and forms, textures, and flavors of the landscape.

The timing of these chanterelles was significant. They would become part of a place-themed welcome feast for my Sustainability Leadership Semester students who would arrive the next day and live for the fall semester at Goshen College's Merry Lea Environmental Learning Center. Seared white-tailed deer chops were also on the menu, from a deer that lived and died here in the headwaters and was processed with the previous year's students. The small portions of meat were accompanied by a medley of vegetables grown onsite and fresh salsa verde. During the semester, the students from Nepal, Canada, Germany, Indiana, and Oregon would have opportunities to learn about bioregional food sources through place studies and direct experience foraging for cattail roots, pawpaws, acorns, mulberries, hackberries, and autumn olive fruit. They would also skin, butcher, and prepare a white-tailed deer as food. These learning experiences combined with interdisciplinary texts by writers ranging from Rachel Carson to Wendell Berry to Robin Wall Kimmerer to Thich Nhat Hanh, would serve as pathways into discourse around place, food, spirituality, ethics, natural history, and other topics that move us closer to the source.

How do we integrate places more intentionally into our lives and our lives more consciously into places? This project began through the editors' friendships and shared experiences around questions like this. We also came to this through mutual care for and commitment to the more-than-human world and involvement in the everyday processes and practices that keep us conscious and aware of the places we inhabit and the beings of all kinds that make our communities rich, particular, and fulfilling.

We are grateful to the many people, places, other-than-human animals, plants, fungi, and others who have contributed in their own unique ways to this book. Our hope is that through the following chapters, the readers will interact with meaningful stories, insightful questions, and authentic perspectives and that the text as a whole might encourage all of us to lean more deeply into our relationships with the more-than-human world. May this book, as Kimmerer suggests, remind us how to live again in a world made of gifts that leave both the spirit and the body full.

Reference

Kimmerer, R. (2013). *Braiding sweetgrass: Indigenous wisdom, scientific knowledge, and the teachings of plants.* Minneapolis: Milkweed Editions.

Goshen, IN, USA   Joel Pontius

Contents

Chapter 1
Hunting, Foraging, and Fishing for Food as Place-Based Learning

Joel B. Pontius

To reach all the way around the elm tree's trunk, my seven-year-old daughter, Emmy, and I have to stretch our arms and fingertips as far as we can. We don't know how long the tree has been standing here, but it is likely that this plant's life began more than two centuries ago. Maybe the elm was present before this land – where my family now lives in the headwaters of Lake Michigan – was stolen from the Potawatomi people in Andrew Jackson's Land Cession 146 of September, 1828. Since then, the deed has included only a few names, as ownership was passed along within one family for many years.

My partner and I heard in perfect timing that the land was for sale, on the edge of the regenerative preserve where I work. Three days after signing the papers at a private equity firm in suburban Fort Wayne, Indiana, a windstorm tore a massive branch from the elm. The force of the branch rippled downwind, pushing a mature red oak and a three wild black cherry trees to the ground. This event continues to cycle through our home in the flames of the wood stove in winter, where these fallen trees have kept us warm. After the early spring rains drench the decomposing elm branch, it also bears some of the season's first wild fruit – the pheasant's back mushroom.

We discovered them, again, last night, on a family amble through the woods. As we are learning to do, we left the largest ones, which were growing from the powdery remains of the largest from the year before. We searched for palm-sized mushrooms, feeling the cool, spongy moisture present throughout them. "They smell like cucumbers," my four year old daughter, Evelyn, reported after taking an audible sniff. "Look, it's my ear!" Emmy said, holding the mushroom on the side of her head. On the steep path to the house, we decided to prepare the mushrooms as a bedtime snack, cut in thin strips and sautéed in butter with mineral salt. Pheasant backs were not the majority of our food this day – there may have been more

J. B. Pontius (✉)
Sustainability and Environmental Education Department, Goshen College, Goshen, IN, USA

© Springer Nature Switzerland AG 2020   1
J. B. Pontius et al. (eds.), *Place-based Learning for the Plate*, Environmental Discourses in Science Education 6,
https://doi.org/10.1007/978-3-030-42814-3_1

calories in the butter – but we experienced them as a gift and as a sign that the wood ducks and round-lobed hepatica would be present soon. As the girls, in their jammies, poked the morsels with toothpicks, the robins – who had recently returned – sang on the wood edge.

About This Book

The main purpose of this book is to consider hunting, foraging, and/or fishing for food as unique forms of place-based learning. The following chapters are written by a range of authors, including individuals who forage and gather, hunt, and/or fish for part of their food, whose lives have been influenced by these practices, and those who are involved in research and advocacy within fishing communities. A number of the authors identify as place-based educators, and write their stories around themes of intentional living as individuals making connections between everyday life, whole food processes, and relating with the more-than-human world. We hope this book will stir the ecological consciousness of readers and inspire fresh trajectories of thought regarding connections between humans, places, food, hunting, foraging, and fishing, and that it might encourage all of us to lean more deeply into our relationships with our places in the more-than-human world.

Narrative Writing and a Search for Truth

The process of writing narratives about our relationships with places creates change – both deepening our relationships with places, and clarifying who, how, and why we are in the world in our particular ways. In Liberating Scholarly Writing, Nash (2004) argues that in the context of teaching and learning, especially, personal narrative writing can be a powerful way to understand, focus, and create meaning in our scholarship and everyday lives. Engaging the personal and social meaning of how we live our lives can help us understand our histories, develop our moral imaginations, and to grow a deeper awareness of how we can shape the health of our lives and those around us. Part of our vision for this book was for authors to craft meaningful, readable narratives that have something unique to teach about hunting, foraging, gathering, gleaning, and/or fishing as forms of place-based learning, and that involve both the writer and the reader in rich learning experiences. The authors used various narrative forms including and across memoir, autoethnography, personal narrative essay, and narrative case study. In the case of narrative writing, truth emerges in the form of trustworthiness, honesty, self-consciousness, livability, adequacy, and other criteria that stem from artful writing (Nash and Bradley 2011).

Place-Based Learning

In this book, we use the term "place-based learning" to conceive of learning in all of its forms – in any context or way of engagement, including connections across cultures, places, species, individuals, communities, generations, and landscapes. The existential power of places open us to new conversations, to connections, disruptions, and nuance that is somewhat difficult to predict (Greenwood and Smith 2014). Place-based learning also allows us to claim our wholeness as ecological, spiritual, emotional beings, and acknowledges the reality that learning is not limited to humans, but includes learning from and beside diverse more-than-human others who are also learning in their own ways (Abram 2010). In a way, as assemblages of diverse human and more-than-human bodies, beings, histories, presents and futures, places, themselves, are learning.

Opportunistic Omnivores

As Paul Shepard (1973) perceptively storied in The Tender Carnivore and the Sacred Game, hunting, foraging, and fishing have played critical roles in building our species and making our human bodies and psyches what they are. From a human ecology perspective, we move as opportunistic omnivores even as our eyes track these symbols that we read on paper or by the glow of the screen, shop at the grocery store, or purchase a used car on the internet. As humans, we have known our places in the world through our diets, identifying ourselves with the plants, animals, and fungi of certain landscapes and watersheds. However, our global systems and technologies have changed so rapidly over time that we have hardly been able to keep up. With more than seven billion people on the planet today, and having reached global carrying capacity for our species in the early 1980s, hunting, foraging, and fishing are not sustainable as ways to feed human society at scale. However, with proper attention to place and commitment to the health of ecosystems and populations first, hunting, foraging, and fishing can be illuminative ways to learn more about the more-than-human, be in our bodies on the landscape, and seek intimate relationships with our food and places.

Eating as a Political Act

For some practitioners, obtaining food through hunting, foraging, and/or fishing has political meaning as a response to this present time of industrial food production in an era of global climate change and ecological instability. Eating is one of the most common things we do, and one of the daily ways we make choices about how we interact with the world's ecosystems. Engaging skillful effort with time and intention, what are the edible aspects of the everyday terrain that might encourage us to look to our yards and common spaces in ways that could seasonally fill our bowls? As the local foods movement has grown, it has become a popular context for diverse people to learn hunting and foraging skills. Tovar Cerulli's (2012) book The Mindful Carnivore, is an example of this trend, and tracks this previous vegetarian's process of learning to hunt white-tailed deer as a way to experience the story of bringing meat to the table while opting out of industrial meat production. Waves continue from Omnivore's Dilemma, Pollen's (2006) comparison of four meals that considered the place, ecological, and health facets of our diets in ways that opened up the contemporary conversation to diverse participants. This book connects with and across the politics of eating and the various processes and practices that bring food to the table.

In This Book

In Chap. 2, Dilafruz Williams takes us on a journey of a one-mile radius where she lives, exposing a rich tapestry of trees, shrubs, herbs, and trailing vines that grace side-walks, city parks, and college campus grounds offering their bounties of food in various shapes, textures, smells, colors, and tastes. She also points to other gifts from her plant relatives: dyes for fabric and medicines for healing and health. Part memoir, she invites us to "notice" with all our senses as she narrates her own experiences of walking and wild foraging in Portland, Oregon, where she has lived for almost thirty years. Paying attention also means being in relationship with birds, insects, and animals with whom these plants are at home. Needing neither a paper map nor a Google portal map, walking the streets, observing, listening, and remembering, she gets to know her place intimately. The web of life of which she is a part, instills deep understanding about relationships with place; no television screen or computer monitor can ever compete with real-life teachings and intricacies of embodied learning. As we walk with her to wild forage, we can feel the bounties of her place: hazelnut, oak, walnut, linden, cedar, douglas fir, pine, apple, cherry, fig, persimmon, plum, aroniaberry, blackberry, marionberry, serviceberry, blueberry, grape, Oregon grape, salal, stinging nettle, rosemary, wild mint, oregano, and fennel, are among the more than one hundred plants that she visits in friendship.

In Chap. 3, Joel Pontius stories his relationship with elk from his first encounter as an eleven-year-old. Join Joel as he learns to communicate with elk, move with them through mountainous landscapes, and as he learns to hunt elk. What other relationships grow in connection with hunting elk? How does his practice of listening change? Through the process, Joel's perspective shifts on what it means to live into place in the more-than-human world.

In Chap. 4, Abbie and Andrew Gascho-Landis join their Alabama friends on their high-fenced hunting ranch, where they encounter their friends' particular cultural traditions from their our own values as a budding ecologist and veterinarian. The ecologist grapples with managing a landscape primarily for economic gain instead of biodiversity and ecosystem health. The once-vegetarian veterinarian struggles with the ethics of eating animals by finding hunting to be more intimate and humane than other meat options. Through two decades of hunting and butchering deer in this managed, yet wild landscape, a new – more nuanced – ethic of land and animal management emerges for them. Now, on their own farm, they work to balance harvest and conservation in their approach to eating meat and tending woods and fields.

In Chap. 5, David Chang and Heesoon Bai present a dialogue on their foraging experiences. David reflects on the impact of shucking oysters on a remote island, catching smelt in a stream and struggling to identify berries from a field guide. Heesoon recalls her botanical education under the tutelage of her mother, who imparted her traditional knowledge on edible weeds, and picking berries with her daughters. Through each of these episodes, they explore the sacramental, cultural, relational, and educational significance of their foraging experiences. Although foraging practices cannot promise to feed the current world population, they suggest that intentional foraging practice can constitute a form of *edible activism*, a way of re-thinking and reshaping participation in a pervasive consumer culture that sees food as commodity rather than communion.

In Chap. 6, Sammy Matsaw expresses how the land of his Oglala Lakota and Shoshone-Bannock ancestors has shaped his practices of hunting, gathering, and fishing. Through these stories he articulates acts of consent and caring for land, water, plants, animals, and human beings. For a larger audience, these stories situate a way of caring and consent that he hopes will begin to help others identity with land and wildlife, so they can feel more at-home in an era of placelessness.

In Chap. 7, Alison Nielson and Rita Sao Marcos, beginning in their own narratives about eating fish, discuss broad issues related to environmental justice for the fishers who struggle to maintain their livelihoods against policies that promote private profit over sustainability. Their stories take them from Canada and continental Portugal to the Azorean Islands in the middle of the Atlantic Ocean where they have been doing research and community work in collaboration with artisanal and small-scale fishers for the past ten years. In discussing history of governance and politics of fisheries in Europe, they outline the struggles for fishing communities. Underneath these stories lie values and images, such as "alive and kicking", that could support the sustainability of oceans and the well-being of fishing communities. Unfortunately, myths and stereotypes about fishers and categorizing industrial scale fishing as the same as that done by people who have deep connections with fish as living neighbors, not dead "resources", are powerful and prevalent. Listening to the voices of fishers tell about living as part of ocean ecosystems while negotiating economic and political systems which champion unlimited growth is a useful way to deal with these complex issues in formal classroom teaching as well as informal and nonformal environmental education.

In Chap. 8, James Farmer delves into the practice of gathering road-kill off the rural back roads and highways of southern Indiana as food. In his detailing of his opportunistic venison harvests, he discusses the history and rise and fall of the white-tailed deer population, as its current abundance lends well to the modern-day gatherer's list of menu items. Farmer examines how gleaning road-kill fits into food security conversation, within the scope of leisure behavior, and ultimately a connection to the land.

In Chap. 9, Amy Azano uses autoethnography and narrative inquiry to examine her father's relationship with hunting as an exercise in actualizing rural literacies to identify opportunities for place-based learning. Azano explores the connection to nature, hunting traditions, the literacy and discourse communities, and the lessons learned from the mountains where her father hunts. The author uses a critical place frame to ask how we might incorporate this culture as meaningful, instructional practices into K-12 schooling. The author takes issue with decontextualized curricula and asks if schooling perpetuates ways of being at risk when it fails to value place and the home knowledge students bring to the classroom. This issue is examined within the narrative of her father's relationship with hunting.

In Chap. 10, Jonathon Schramm expresses how neighborhood foraging – drawing calories and spiritual connection from creatures whose everyday familiarity can cause us to overlook them – can be an important way to reconnect modern urban residents to the mysteries of the natural world. In this essay, he explores some of the practical skills that he has acquired to harvest maple syrup and edible mushrooms from a sugar maple tree (Acer saccharum) that graces his backyard. Through that work, he has enjoyed opportunities to interact with and teach his neighbors about the potential for wild eating from a typical neighborhood setting. Even more fundamentally, he explains how he has seen his own connection to the natural world greatly change, in ways that his formal ecological training did not always convey.

In Chap. 11, focusing on a community scale, Alun Morgan, Emma Sheehan, Adam Rees, and Amy Cartwright discuss the opportunities and challenges of collaborative and participatory work across a range of stakeholders groups – scientists, fishers, and the local community – to address issues of social, environmental and economic sustainability within coastal fishing communities. Using key concepts in 'social learning for sustainability', the particular case study of the Lyme Bay Marine Protected Area highlights key themes of 'social learning' undertaken by the various stakeholder groups. Personal narratives by key personnel engaged in a particular Citizen Science research initiative provides insight into the personal and professional motivations and learning outcomes from engaging in such work, as well as broader societal impacts.

Finally, in Chap. 12, Mike Mueller reflects on how experiences of despair and also hopefulness helped him breach the boundaries of what is known as *food harvest*. His chapter ends with a heuristic that is written to help start conversations around the politics of food consumerism.

About Hunting, Foraging, and Fishing

In this book, we position hunting, foraging, and fishing for food as practices to be used carefully in response to the particular places we inhabit, personal values, and ethical perspectives. At this point in the history of the planet, hunting, foraging, and fishing are not scalable solutions to global food systems issues, and may not be viable options in many places and cultures. However, practiced in ways that respond to the ecological needs of particular places, contribute to the personal and/or spiritual meaning of connection with the more-than-human, and provide nutrient-dense local food for communities, hunting, foraging, and fishing for food may have beneficial outcomes on multiple planes. In the following, we provide a brief frame for each of these practices.

Hunting

The term "hunting" applies to a wide set of practices, from the stereotyped high-fenced trophy hunters who pay to kill large-antlered animals that have no way of escaping, to indigenous hunter-gatherers like the Yanomani people of the Amazon River Basin who rely on meat from animals as part of their daily diet, to those who participate in hunting as a seasonal cultural tradition. Considering the wide continuum of cultural values, practices, and landscapes that bring hunting into form, it is most accurately understood in the embodied context of particular places, culture and time (Pontius et al. 2013). Our initial purposes for hunting were for food to eat, and some still put this purpose first. However, as human cultures and the world have rapidly changed over the past 10,000 years, the ways that people hunt – and in many cases the reasons for hunting and meaning derived from the practice – have taken radically different form. Through the industrialization of the world, hunting practices have changed substantially for most cultures. The development of vehicles, airplanes, land navigation, weapon technologies, drones, optics, automated cameras, and a wild list of other gadgets have led to critical questions about what it means to be ethically responsible within the more-than-human landscapes where hunting takes place. Aldo Leopold (1949) was asking similar questions around the ethics of hunting gadgetry in the 1940s, responding to technological advances in optics and the accuracy of rifles. Here we are, less than a century later, with legislation in the works to put limits on searching for wild animals with long distance drones. Practiced in particular circumstances where hunting contributes to the health of animal populations and ecosystems, provides nutrient-dense local food for the table, and develops deeper relationships between humans and the more-than-human world, it can be an illuminative practice for place-based learning. For this to be the case, we will need to draw on a wider ecology of stories and storytelling that engages deeply with ethics and the ways people live with other beings in place (Russell 2019).

Foraging and Gathering

From the foraging practices of people without stable housing, to the gleaning of local fruit trees by local ENGOs to support the health of people in poverty, to the hunting of mushrooms deemed exotic by diners at expensive restaurants, foraging food from the local landscape is being explored as an economic, nutritional, spiritual, recreational, and culinary opportunity. From the perspectives of urban planning, environmental regulation, and socioeconomic inequality, these newer cultural practices are also often rife with political conflict. Indigenous groups continue the seasonal gathering traditions of wild foods that predate European colonization and settlement. Contemporary political contexts, such as the conflicts surrounding the wild rice harvests in northern Minnesota, demonstrate the legacy of colonialism and its impacts on traditional food practices for people worldwide. The overharvesting of wild leeks has left populations struggling or locally extinct, which shows that protections must be put in place and enforced. In Europe and the United States, foraging is regaining popularity as a way to reconnect the body with the land and meaningful, nutritious food. "Wildcrafting" is a popular trend on social media, where individuals share photos of foraging greens, mushrooms, berries, and materials for making baskets. Foraging with restraint – and in ways that are responsive to the health of plant communities – can steep practitioners in the phenology of local landscapes, provide healthy food, and allow place-making to occur authentically in our everyday terrain.

Fishing

Many of the current discussions around fishing and sustainability are focused on depleted and threatened ocean fisheries, brought about through a whole slew of ecological issues: overfishing, climate change variables such as ocean acidification, rapidly warming ocean water temperatures, the ocean plastics crisis, and others. Commercial/industrial markets for wild fish continue to put heavy pressure on ocean fisheries for food that is imported and exported globally. The impacts of failing fisheries can be devastating for indigenous and nonindigenous communities who depend on waterways to meet their needs in places such as Alaska, USA, where residents continue to rely on salmon populations for food in ways that connect across the social, economic, ecological, and spiritual aspects of living in places (Lord 2016). While fishing can be harmful to populations at larger scales that may not be responsive to place, local communities are responding through advocacy, school citizen science and youth free-choice educational programs that include culinary projects, engineering and green business, habitat restoration, dam removal, and activism that may strengthen fish populations and food security.

Join Us

We hope reading this book is like sitting down for a meal that we foraged, hunted, fished, and prepared together. As you read, pause to talk with David Chang and Heesoon Bai as they shuck purple-fringed oysters into a metal pail. Taste a spoonful of plum and marionberry puree' before Dilafruz Williams starts the dehydrator, and sample the finished maple syrup that will be drizzled over pears while Jonathon Schramm reads the hydrometer. Make venison meatballs with Abbie and Andrew Gascho-Landis, and give Alison Neilson and Rita São Marcos a hand preparing bacalhau (dried cod). Coat the pepper steaks with Amy Azano while her father, Donnie Price, tells the story of his deer hunt at Cedar Creek. Pass by the stovetop, where Sammy Matsaw stirs magenta wozapi – traditional Lakota chokecherry porridge – and shares stories about spear-hunting for Chinook salmon.

Join us as we celebrate the richness of places, the company of each other, and the wisdom of the more-than-human world. Let us sit down together and give thanks.

References

Abram, D. (2010). *Becoming animal: An earthly cosmology*. New York: Pantheon Books.

Cerulli, T. (2012). *The mindful carnivore: A vegetarian's hunt for sustenance*. New York: Pegasus Books.

Greenwood, D. A., & Smith, G. A. (2014). *Place-based education in the global age: Local diversity*. New York: Routledge.

Leopold, A. (1949). *A sand county almanac, and sketches here and there*. New York: Oxford University Press.

Lord, N. (Ed.). (2016). *Made of salmon: Alaska stories from the salmon project*. Fairbanks: University of Alaska Press.

Nash, R. (2004). *Liberating scholarly writing: The power of personal narrative*. New York: Teachers College Press.

Nash, R., & Bradley, D. L. (2011). *Me-search and re-search: A guide for writing scholarly personal narrative manuscripts*. Information Age Publishing.

Pollen, M. (2006). *The Omnivore's dilemma: A natural history of four meals*. New York: The Penguin Press.

Pontius, J. B., Greenwood, D. A., Ryan, J. L., & Greenwood, E. A. (2013). Hunting for ecological learning. *Canadian Journal of Environmental Education, 18*, 80–95.

Russell, J. (2019). Telling better stories: Toward critical, place-based, and multispecies narrative pedagogies in hunting and fishing cultures. *The Journal of Environmental Education*. https://doi.org/10.1080/00958964.2019.1641064.

Shepard, P. (1973). *The tender carnivore and the sacred game*. Athens: The University of Georgia Press.

Chapter 2
Urban Wild Foraging–Walk with Me, a One-Mile Radius

Dilafruz R. Williams

….the place to observe nature is where you are; the walk to take to-day is the walk you took yesterday. You will not find just the same things: both the observer and the observed have changed. (John Burroughs)

I have lived in the same neighborhood in the Johnson Creek Watershed in Portland, Oregon, for 27 years. I know my place intimately as I have walked at least along a one-mile radius from my home hundreds of times. The walks have awakened me to a deep understanding of my connection with this place I call home as I continue to become conscious of and notice the intricacies of both the *life-less* and the *life-bearing* contrasts that my place reflects.

First, the *life-less*: Concrete, asphalt, brick, glass, metal, are a common sight that greet me when I take a walk. The streets are grey and asphalted; the cross-walks are white-striped; the homes and fences are of varying shapes and sizes; the street signs are newly painted every few years for readability; the once unpaved roads are paved and modernized to support increasing numbers of vehicles; the housing lots are split to construct ever-more new homes to accommodate the growing population moving to Portland; and the roads are excavated to replace the hidden sewage pipes with modern more durable ones. The latter has been a relatively recent multi-year project undertaken by the City government. It has entailed trucking in colossal equipment and machinery for cracking and breaking the concrete and digging over twenty feet of earth through the nearly-impermeable asphalt to remove the corroded and outdated sewage pipes. These have been replaced with modern and considered to be longer-lasting cylinders in order to carry our "out-of-sight, out-of-mind" bodily excreta and waste daily from homes to wastewater treatment plants and prevent the sewage overflow in the river. This excavation project to modernize sewage pipes also exposed the hitherto buried and unseen gigantic roots of trees intertwined across species as if holding hands underground and supporting one another in defiance of the machines' damaging onslaught and indiscriminate chiseling, hammering, slicing, and bulldozing.

D. R. Williams (✉)
College of Education, Portland State University, Portland, OR, USA
e-mail: williamsdi@pdx.edu

© Springer Nature Switzerland AG 2020
J. B. Pontius et al. (eds.), *Place-based Learning for the Plate*, Environmental Discourses in Science Education 6,
https://doi.org/10.1007/978-3-030-42814-3_2

If it were not for these roots anchoring the trees and shrubs that grow above ground despite modern technologies that bring destruction to them for convenience to humans, my neighborhood would probably not be much different from other American urban neighborhoods in many respects. That is, the life-less icons and symbols of cities are duplicated everywhere across the United States. Beyond these canons of mundane urban ventures, what makes my place, my neighborhood, distinctive are the myriad living entities that are unique and priceless in their abundance. Walking one block at a time, we meet trees, shrubs, vines, herbs, and "weeds" along with the wildlife they attract: *Life-giving*, alive, magical, biologically diverse. We also meet human neighbors of all ages. Here, I present some reflections on my relationship-building adventures especially with the life-embracing plants that grace my neighborhood. The bounty, abundance, and variety of plant offerings make the place where I live truly distinctive. I marvel at each new encounter with plants and the associated wildlife and kinship that I develop with them: I meet. I wonder. I learn. I forage. I wonder. I learn. Wild foraging helps sustain my on-going, persistent, delicate relationship with place.

My neighborhood offers much that I do not or cannot personally grow as food, despite the fact that I am a gardener. I follow five principles when I forage: (a) Gather only that which grows in parking strips, public spaces and parks if allowed; (b) Look for signs of spraying – e.g. is a yard well-manicured with no weeds at all? That would likely be a sign of spraying; do not pick in the vicinity when in doubt; (c) Ask permission of property owners or tenants when foraging from their property; (d) Take only what I need, and primarily that which has dropped; rule of thumb is, less than 0.5%, unless there are signs that no one else is gathering and fallen fruits and flowers will likely rot; and (e) Reciprocate; ensure that others can forage from my own place by planting foods in the parking strip where I live. Generosity involves both receiving and giving.

Gathering food requires us to leave the car and even the bicycle and get into a habit of walking. Similar to Frederic Gros (2014), I do not view a walk as a means to get from "here to there." Most of the time it is undertaken with an open mind and without any goals. Each step is a celebration toward accepting life in all of its intricacies and mysteries. I have found that walks always offer surprises and new understandings despite the same encounters. Foraging involves a commitment to become an "inhabitant" such that neither a google portal map nor a paper map are necessary to know where I am. After all, a map is not the territory. The diverse plants themselves are incredible markers of street signs that also help with sharpening my memory of location. David Orr's (1992) distinction between an inhabitant and a resident is insightful:

> A resident is a temporary occupant, putting down few roots and investigating little, knowing little, and perhaps caring little for the immediate locale beyond its ability to gratify. …The inhabitant, in contrast, dwells…in an intimate, organic, and mutually nurturing relationship with a place. (p. 130)

Beyond residency, I have chosen to dwell. Or perhaps the plants, the wildlife, the neighbors, and this watershed, have inadvertently chosen me to dwell, to help me to become attached to my place in profound ways. I invite you to join me on my journey and experiences of discovering some of the food and healing/medicinal plants that I harvest during walks in my neighborhood, a one-mile radius with my house on Rex Street at the center of the circle.

For me foraging is not about saving money on foods, though it is for many foragers. In my walks I have developed a sense of place and belonging as I have become attached to the ongoing relationships with life where I live. Perhaps, as an uprooted immigrant, I have had an unconscious need for inhabitation. And, perhaps, as a botanist and a lover of plants I am simply drawn to them. I have met, and built relationships with, at least one hundred varieties of trees, shrubs, herbs, and vines that we would encounter, walking in all directions. I anticipate that the walk would spread over several days and seasons. Given the limits of our walk together, I am unable to show you everything. However, here are a few friends that give a flavor of the abundance and biodiversity that bring richness to my place. Trees: walnut, linden, hazelnut, oak, crabapple, cherry, plum, red cedar, chestnut, douglas fir, pear, fig, mulberry, pine, persimmon, hawthorn. Shrubs and vines: blackberry, marionberry, raspberry, serviceberry, grape, salal, and Oregon grape. And Herbs: rosemary, oregano, sage, wild mint, fennel, lavender, lemon balm. I have discovered over time that each has a unique contour and configuration of branches, barks, leaves, flowers, and fruits or cones. Each has a unique relationship with wildlife. To note the nuances and to understand these in deep ways, I have learned to transcend the obvious. I will share in depth one wild foraging relationship – with the walnut – and touch upon several others in this essay.

To appreciate the foraging process, we need to understand intention. Plants are more than food producing entities for the benefit of others. They are in relationship with other plants, wildlife, humans, and with living soil. Various species of spiders, bees, wasps, ladybugs, earthworms, slugs, snails, and more interact with the plants. And so do the finches, jays, flickers, wrens, hummingbirds, chickadees, swifts, robins, crows, sparrows, starlings, and egrets and ospreys by the streams, along with squirrels, moles, and more. Soil, too, surfaces in human bodies and consciousness. With every bite of food, I nibble at soil's history, biology, and culture, even as I ingest its nutrients. Soil's sensorial memory is entwined with my bodily memory. By virtue of its life-giving and life-nourishing qualities, soil serves not merely as a medium in which life grows. Soil is life. It is living (Williams and Brown 2011, pp. 43–44).

As Michael Pollan (2001) explains in the *Botany of Desire*, we humans do not stand outside the web of nature, we are an integral part of it. When gathering in the wild, we do not need to exclusively focus on food for feeding, as plants are also believed to have capacities and properties for healing. In many cultures such as mine (I was born and raised in India) food and medicine are not dichotomized. Many foods *qua* foods reflect their healing qualities and assets. They are also used in rituals and ceremonies. In keeping with these holistic traditions of integration of ceremonies that I was raised with, and viewing plants as healers, when I forage I also look for ways to support health as what I gather serves multiple purposes for the well-being of body and mind. For example, seasonal tea-making is one such ritual that foraging has supported. Further, seasonal berries can be preserved by making jams and jellies. Fruit rolls are another form of delightful preservation. The antioxidants are an added bonus. Thus, after gathering, I also preserve the food and invite my friends, students, family, and neighbors to join me in preserving. For many, it is their first time. While not covered here, in wild foraging, I also make dyes for cloth and crafts such as holiday wreaths from dried grape vines. Innumerable plant craft workshops are available in the community. Even though this writing expresses my personal experience, with "first person singular" being used freely, in reality foraging would be a lonely enterprise were it not for family and friends either walking the one-mile-radius with me or helping to preserve the bounty. I also believe that quiet walks alone are necessary and to be prized as the senses get finely tuned and sensitized in remarkably different ways.

Some days I have walked only one block, pausing many times. Like peach farmer David Masumoto (2003) who considers his "art of seeing" as a critical capacity in knowing his peaches well, I, too, stop, look around and upwards, not just down at the grounds, so that I can observe the plants and listen with care. I also encounter children and adults during my wild foraging strolls. Curiosity and wonder are a good start. A nod, a hello, introductions, and checking-in can go a long way in building relationships. Some lead to new friendships; others to becoming community. Walking the same paths, I get to know not only the names of the streets and the variety of design configurations of the houses, but I also recognize the biodiversity of plants with their mystical scents, their myriad shades of green, their countless shapes and textures, along with the magnificence of the birds, animals, and insects that flock the flora. And, more importantly, I know the *seasonal* signs of the gifts I am likely to receive from the plants as I await in anticipation to forage; with each month there is a new expectancy.

Consider walnuts. In the block where I live, there is a striking fifty-foot tree that attracts birds, nests them, and, around September and October, drops the walnuts. The tree is oblivious to where the fruits fall: asphalt, a parked car, sidewalk, or manicured grass lawn.

On an early autumn day, over twenty years ago, I stepped on and almost slipped on the husk of a walnut little realizing what it was until I looked at the dozens of messy "balls" of cracked green fruit spread over the sidewalk and the streets. I looked up at the tree. An amazingly staunch tree with green fruits left me in awe. Within a span of ten minutes, I had assessed the number of walnuts that had already fallen and the myriads that were still holding on to the tree. Foraging started: first, with the ring of a house bell to take permission from the owner where the tree was growing. Not interested in harvesting, and grateful that I was going to clear up the "mess," the elderly woman gave me permission to forage and a paper bag to gather the fallen fruits. Puzzled and pleased, my walnut gathering adventure began. Next, I brought my son along as we walked a few houses down the block with a bag each in order to pick up more of the rich fallen fruits, several with their husks torn open. It's messy, but walnuts are best picked after they fall as that is when the fruits are ready. The lure of the fruit led us back to the tree every couple of days. Not all fruits that we collected were useable as food, since some had worms, fungus, and rot. I invited a close friend, a native of Portland, who loves nuts. Together we gathered the bounties over a couple of weeks. One neighbor stopped by out of curiosity but did not want to touch the split fruits due to the dark "ink" of the husk.

Walnuts are complex fruits. They come down with a tough green husk, also known as hull. These husks are rather soft, unlike the inner hard-shelled walnuts that we might see in a store (if we buy "whole" walnuts). The husk can be pulled apart exposing the brown hard shell. A walnut's tannin stains hands when the green husks are peeled off. The first time I foraged, I did not realize this; I learned through the experience of ink-stained hands and clothes to subsequently use gloves. After the husks are removed, the fruits, now with hard brown shells, have to be washed and then dried. I have had success in drying them outdoors, though some people dry them in the oven. Once dry, my family shells them by hammering and separating the nutmeat or kernel with a picking tool. Next, we store the freshly uncovered walnut in glass bottles. Refrigerating can preserve them for up to two years.

Foraging involves being present somewhere at the right time, and being in the present while gathering free food and watching the interconnected dance of plants and place. Moreover, foraging can also evoke memories. When I forage walnuts, for instance, I recall my childhood experiences of competing with my brothers. Who could shell the whole, almost-spherical, hard walnut with a hammer or stone with one blow that would crack it open into perfect halves? I grew up in India at a time when there were no pre-packaged foods or supermarkets. Unlike our present-day chopped walnuts sold in packages, they were bought whole and intact, with their hard shells, at a special mart where my parents had a relationship with the owner who sold only walnuts; his was an intergenerational small business. We were intergenerational customers. There is something so "earthy" about the color and texture of the hard brown shells of the walnut. As siblings, we competed to crack it open, as we were intrigued by the perfect half of a walnut exposing the contours and patterns resembling the brain. Folklore, in many traditions, including mine, conveys that since walnuts resemble the brain, eating them strengthens our memories.

My relationship with the walnut tree continues. A surprise encounter over twenty years ago has brought me to this tree over and over again. Foraging started serendipitously, with the loss of physical balance, a slight slip on the fallen fruit. But when the first rains arrive in autumn, the trees' gifts right my balance and revive my memories.

An interesting phenomenon I have discovered over the years is how much biodiversity exists within a circle of a one-mile radius. And, visiting and revisiting the same plants over time, I have felt that each has a unique personality. Each embodies a story. For example, I distinctly remember feeling intoxicated with the aroma of a tree as I came close to it on a May morning almost a decade ago. I looked up at the tree, which I later discovered is the Linden, also known as *Tilia,* or basswood, in the northwest. It has a sturdy trunk with innumerable braches that divide and subdivide. The light yellow cascading flowers had attracted hundreds of bees that were buzzing in the tree, much as I was attracted to both the beauty and the perfume of the flowers. The leaves of linden, unlike that of the walnut, can also be added to salad. While the leaves, too, have health benefits, I forage the flowers that fall by the hundreds each day. I steep the flowers, dry or fresh, in hot water to make tea. It is believed that the tea can help with colds, coughs, fever, and high blood pressure. Linden also helps with detoxifying the body. Since linden tea has a calming effect, it is used as an aid for anxiety and stress and to support general well-being. I dry the flowers directly in the sun and store them in a glass jar. When possible, I barter these fortunes with other herbalists in the community. Basswood honey is prized in my region. Linden's folklore and myths are to be valued, too. In many cultures, linden is considered to be a holy tree. In German mythology, linden represented peace and justice. Not only did people dance under the linden tree but judicial meetings were held in its presence as the tree was believed to have the power to uncover the truth. Other European cultures, too, have considered the linden as a sacred tree. Linden, thus, graces us with its many gifts.

Ceremonies and rituals that include making tea which is healing and soothing, or making jam, jelly, and fruit rolls, or drying and preserving parts of the plants that have medicinal value, are also part of the cycle of foraging beyond gathering and eating the fruits raw. Seasonal abundance makes us mindful of how to feed and heal the body beyond the specific months when the plants visit us.

One of my family's prized activities is preserving berries after harvesting; this helps with completing the cycle of foraging. I have found blackberries, raspberries, currants, and marionberries on my walks since there are a number of public parks and spaces where they grow in abundance. Each year, young and old forage blackberries from Springwater Corridor, a multi-use trail that runs parallel to Johnson Creek, one of the last free-flowing streams in the Portland region. In her poem, "August," Mary Oliver (1993) captures my sentiments when she writes how she cannot stop reaching for the blackberries in the woods and eating them. Recognizing her connection to the creek and the honey enticing flowers, she claims: "there is this happy tongue" (p. 143). To stretch culinary happiness beyond summers, I often make jams and fruit rolls with others, experimenting each season with new combinations of berries and spices. See Fig. 2.1 for my recipe.

I distribute the chewy fruit rolls as gifts to many friends and kin. The vibrant colors, unique texture, divine taste and aroma, feed the soul. Teenage children of my friends and neighbors and my own nephews and nieces willingly offer to help to make fruit rolls as they simply cannot resist its taste, once they have had a bite! Figure 2.2 presents a flavor of a conversation with teenagers Victor and Gabriella, as we make fruit roll.

> …And then, the next day, the two teenagers come back, with two friends. They are thrilled to show off and share their delicious fruit rolls. Their making. And, it starts all over again.
>
> Victor: *Can we make more, please?*
> Dilafruz: If you can pick the berries.
> Gabriella: *How about cherries? There's a cherry tree on my street and people step on all the cherries. Can we make cherry fruit rolls?*

Thus starts intergenerational learning: foraging, cooking, experimenting, tasting. In order to share the joys of foraging with others, I have planted a number of fruit-bearing shrubs where I live: aronia, marionberry, blueberry, currant (red, black, yellow), mulberry, raspberry, strawberry, Oregon grape, salal. In fact, there is berry-abundance at my own home; hence I invite neighbors to forage from my

---

Plum-Marionberry Fruit Roll Preservation

Ingredients: Plums as base ingredient, marionberries (can be replaced with strawberries, raspberries, blackberries); pectin without sugar, lemon, ginger

I wash the plums and place them in colander to let the water drain off. Next, I take off the pits from plums, and, leaving the skin on, I chop up plums into small pieces.

For four cups of chopped plums, I add a packet of pectin without sugar; if plums are too tart then I add half a cup of sugar. After that, I add half the amount of the scrumptious marionberries, which have also been washed and strained. I marvel at the colors—purplish red plums and maroonish purple berries. I cook on gentle heat until it turns to a jelly-like resilient consistency. The color often changes to a deep maroon, and a heavenly aroma of berries starts filling the house. Next, I add one-eighth of a cup of freshly squeezed lemon juice (no seeds), and then a spoonful of freshly ground ginger. Ginger is a rhizome that I am fond of since, in India, we use it for much of our meal-making given its medicinal value for supporting digestion. Stirring all ingredients well on slow fire, the fruit is ready if the mixture is thick enough when poured from an ice-cold spoon and does not run like juice.

In the meantime, I ready the trays from the dehydrator by spraying them with olive oil to prevent the fruit from sticking onto the trays. I pour the puree onto the trays, and smoothen the surface with spatula. Four trays are stacked in the dehydrator operating at 135^{0}F, for at least 6-7 hours. Dehydration requires patience. The fruit roll is ready when it peels off the trays easily. Finally, the fruit roll is transferred from each tray onto parchment paper, and is either cut up into strips or rolled up.

Fig. 2.1 Recipe: plum and marionberry fruit roll

parking strip. I have been fascinated with the knowledge that berries are rich in antioxidants boosting our immune system.

Our palates are so modernized with salt and sugar added to foods that anything bitter is not welcome. Yet, a fair number of bitter plants have medicinal properties that help with a variety of ailments including indigestion. Oregon grape is one example of a quintessentially bitter plant that has the capacity to cleanse the body. It also has the honor of being the state flower. It is a prickly, evergreen shrub. While it does not belong to the grape family, its clustered fruits are purplish-blue and look like grapes. It grows prolifically in my neighborhood and can be seen as a fairly dense 5–6 feet hedge in a number of areas, including parks and the college campus close by. Its leaves, densely packed on stems, remind me of holly with its sharp-toothed leaves that are also leathery and dark green. In summer, the yellow flowers turn to berries that hang on the plant for several weeks. When they are uniformly dark, the berries are ready to be picked; I slide my fingers down each bunch as they fall into my basket. Oregon grape is sought by herbalists given that it is very rich in antioxidants and except for the leaves, all parts of the plant – its roots, bark, flowers, and fruits – have medicinal value. We can scrape its bark (and even the roots) to

Intergenerational cooking

Victor: *Why do you add the berries to the plums?*

Dilafruz: Don't have to, but the taste can change and it is worth the variety.

Gabriella: *Wow, look at the color. It is changing with the heat!*

Victor: *Oh wow, the skin is melting away.*

Gabriela: *I had no idea. When my mom cooks tomatoes the skin also starts sort of becoming soft.*

Victor: *How long do we keep stirring?*

Dilafruz: Until it no longer drips from an ice-cold spoon.

Gabriella: *Is it ready now?* Tries to pour it from a spoon to check.

Gabriela*: Look it is getting thicker.*

Victor: *Yes. And it smells so good, just like it tastes. Can't wait.*

A few minutes later, as trays are being prepared:

Gabriela: *Why are we spraying olive oil on the tray?*

Victor: *Can I try?*

Dilafruz: Why, do *you* think? In what ways do you think it helps?

Victor: *Stop it from sticking?*

Gabriela: *Yeah*!

And, finally, as the lid goes onto the dehydrator:

Victor: *How long will it take?*

Dilafruz: About six to seven hours, maybe more.

Gabriela and Victor together: *Oh no! We have to wait that long?*

Dilafruz: Okay let's start cleaning up.

Fig. 2.2 Intergenerational cooking, a dialogue

expose the yellow alkaloid, *berberine*. I have taken classes with an herbalist who specialized in this plant, and subsequently planted it where I live. I learned the right techniques to forage and preserve the various parts of the plant, especially the stem. While roots can also be foraged for the same purpose, I do not like to uproot plants when I forage. If a stem is gathered, then with a sharp knife I peel its outer bark; when dried, it can be stored in a glass bottle to be used for immune enhancement and treat a variety of gastrointestinal ailments.

The significance of ceremony and rituals among indigenous populations across the globe is well-documented. In several cultures and countries, the tea ceremony is practiced as a ritual to foster a variety of outcomes including the promotion of harmony with nature. Along with food preservation, tea-making is another event that makes foraging memorable. A close friend and colleague Judy Bluehorse Skelton, who is a Native American herbalist, makes a ceremonial tea for my students in the *Sense of Place* class that I teach in the Fall term each year. For those students who are new to Portland, this course helps them become grounded in the program and in thinking and wondering about "place". We hold the class at the Learning Gardens Laboratory (LGL), a 13-acre property opposite a public school in an economically poor neighborhood which otherwise seems rich in plants worth foraging. Within walking distance of my house, LGL, which I co-founded in 2004, is an educational and training center for garden-based learning. In keeping with Native American traditions, Judy takes us through the ritual of first giving thanks and expressing gratitude to the plants, the water, the wildlife, the air, and the soil as we forage. Respect for and cultivation of relationships are stressed as gathering begins (Bluehorse Skelton 2008). Judy uses plants that we forage at LGL where the class is held: a few sprigs of Western Red Cedar, a short branch of Douglas Fir including the needles, and some fennel seeds are added to a pot of boiling water. She lets it simmer for about 15–20 minutes while we hear stories and get a lesson on the health-giving and medicinal qualities of these plants for curing coughs and colds. The aroma of the plants fills the greenhouse where we usually gather around tables that are set up. Letting the tea steep for another 20–30 minutes, students share their stories of place-making especially associated with smells and aroma that can bring back memories of their childhood places: a grandmother's pumpkin pie, a mother's steaming soups, or an aunt's baked gingerbread cookies. As we sip the tea that is served, our palates are sensitized and our memories are revived. In many cultures, people sit at a table sipping tea while enjoying conversations even as it serves as a healing drink. Students in the *Sense of Place* class begin their journey toward place-making through this simple process of foraging and tea-making that have profound significance during the course of their academic program of study.

Foraging requires reciprocity. Not only do I forage as I walk the one-mile radius, I also ensure that I offer others in the neighborhood opportunities to forage where I live. Hence, the parking strip and front yard on 42nd and Rex Street have blueberries, raspberries, currants (black and red), serviceberries, oak and more. Over the past quarter century, plants have birthed or voluntarily visited, grown, and withered. The cycle of birth, decay, death, and rebirth continues. These plants are available for neighbors and wildlife even as they are for me and my family. It is one way to meet people including children in the neighborhood. Three summers ago, there was a knock on my door. About a 4-year old boy asked my permission to pick the blueberries in the parking strip while his father stood at the bottom of the concrete stairs. Providing a small container to him, I showed him how to pick only the ripe blueberries with care, and also explained that he should pick just "enough" and leave the fruits also for others, including the birds, to eat.

Urban Wild Foraging: Exotic Trend or Necessity?

Is urban foraging an exotic trend that will pass or is it a necessity spreading globally as the public discovers an opportunity not only for free food but also for belongingness, community, and identity resulting from knowing one's place intimately? For a rather small, though increasing, number of people, it has become a necessity. The turmoil of transnational diaspora, refugee movements, and extremes of climate change including desertification are factors that create a sense of urgency and necessity for those in search of good healthy food while simultaneously in search of connections to place and communities. For decades we have been mesmerized by the golden promises of the conveniences of all that is artificial, and yet there's a catch: The vehicles that serve as our alternative limbs carrying us long distances away from our places and neighborhoods, atrophy our leg muscles and bodies, and disconnect us from where we live. The never-ending upgrades of appliances and machines promise leisure yet to be found. The cellphones keep us connected with distant others in space but not place. Even the microwaveable "food" some of which has travelled 5000 miles to make it to our guts, while expediently picked up at a grocery store is mass-produced, packaged, and becomes a "commodity" of convenience. The list is endless. Uprooted, dislocated, and for even those who have a home, people are on the constant move. Many of us do not know our next door human neighbors, nor are we curious about the fauna and flora – the *non*human neighbors – that inhabit our places, our watersheds, and our neighborhoods and with whom we share the urban locale.

Within this modern reality, foraging provides a cautious hope. There are signs that perhaps with the turn of the century, the twenty-first has been a time of waking up, of remembering the hitherto lost and forgotten, with plants inviting us back into relationship with *life* and its offerings of magnanimity, awe, wonder, and grace, right where we live. One such effort is about growing food locally, eating healthy, and being in community sharing the bounties offered by land; community-plots in many cities attract so much interest that there are long wait-lists. Another effort that I have shared at length – urban wild foraging – is flourishing. A search for "urban foraging" on google.com presents us with a plethora of books and "how to" manuals for this relatively recent trend. A sample of sources is presented in a separate section of the References. Technical advice, specific to locale, is available on foraging: how to gather, scavenge, process, and consume edible wild plants and other foragables. Formal tours and guides are provided locally by some of the "experts" to encourage people and to teach them how to glean food from public spaces. Foraging experts take novices on urban walks to explore the nooks and corners and the public spaces of their city in search of food: parking strips, side-walks, parks, abandoned lots, and city forests. Young and old join these walks to learn to identify and gather foods. Drawn to these ventures, many want to figure out how to minimize their reliance on the energy grid for food. They have taken to the bounty of foraging the streets where they live.

For a sample of the resources, a number of which have self-explanatory titles, see the *References* at the end of the chapter: "A Sample of Sources and Google Sites on

Field Guides and Wild Foraging." (E.g. Brill 2012, Brill and Dean 1994, Craft 2010, Lerner 2013, Thayer 2003, Vorderbruggen 2016). Several guides to the edible wild plant world are available that include how to identify them, the ideal times to harvest them, and a variety of culinary uses. Some include cultural information about rituals associated with plants. There is also an interactive website, "Falling Fruits," (https://fallingfruit.org/) launched by Caleb Phillips and Ethan Welty (2013). This website has close to a 2000 edibles spread over 1.2 million locations, represented by dots on the map, across the globe. The website is an interactive space with information provided by citizens, and from databases which are hosted by governments and associations that promote foraging and gardening in various locations. One can zoom in on the dots on the map for a description of what is available in specific locales. Furthermore, *Wild Apps* for foraging are designed for both i-phones and i-pads; these allow people to freely explore edible plants in their neighborhoods. *Foraging Flashcards* are also accessible on i-phone apps that support the identification of edible plants.

An annotated bibliography and review of literature conducted by Rebecca McLain and her colleagues about human-plant interactions and foraging non-timber forest products is one of the few concise publications covering 140 research articles, field guides, websites, and articles in the popular media as there are very few studies of urban gathering (McLain et al. 2012). It documents that "gathering persists and may be growing in U.S. cities, is of interest to people of many walks of life, and provides a range of benefits to gatherers and the communities in which they live" (McLain et al. 2012, p. 2). Among the findings, they conclude: "…people derive numerous benefits from gathering plants and fungi in U.S. cities. Gathering provides useful products, encourages physical activity, offers opportunities to connect with and learn about nature, helps strengthen social ties and cultural identities, and, in some contexts, can serve as a strategic tool for ecological restoration" (p. 5). Analyzing five studies on urban foraging conducted in Baltimore, New York City, Seattle, and Philadelphia, Rebecca McLain and her co-authors conclude that foraging is a vibrant and ongoing practice among diverse communities (McLain et al. 2014). The number of foragers interviewed in the studies ranged from 8 in Philadelphia to 58 in Seattle. According to the researchers, foraging not only supports livelihoods, it is also a means for acquiring essential foods, medicine and materials for households. There are also opportunities for the local residents to directly connect with nature and develop new social relationships. While government policies often discourage foraging, they conclude that "…foraging can be seen as a deeply relational practice connecting humans with nature, other humans, and their inner selves," (McLain et al. 2014, p. 231).

The connections between foraging and cultural belonging were also made by foragers who self-identified as newcomers or immigrants. In a study of urban foragers in Seattle, Melissa Poe and her co-authors (2014), similarly found that:

[m]any foragers spoke of how through foraging they had created a deeper, more intimate knowledge of the city with layered meanings built over time…Active relating, moving, and engaging (not simply being) with plants, mushrooms, and spaces in the city were therefore processes though with foragers came to belong. (p. 901)

This sense of belonging is often tied to establishing roots, knowing one's place. Place identity calls into play our affective faculties, and "answers the questions: Who am I?-by countering-Where am I? or Where do I belong? From a social psychological perspective, place identities are thought to arise because places, as bounded locales imbued with personal, social, and cultural meanings, provide a significant framework in which identity is constructed, maintained, and transformed," according to Lee Cuba and David Hummon (1993, p. 112). But, thoughtfulness about our relationship with place is in order. As Wendell Berry (1991) urges:

> No place is to be learned like a textbook or a course in a school, and then turned away from forever on the assumption that one's knowledge is complete. What is to be known about it is without limit, and it is endlessly changing. Knowing it is therefore like breathing: it can happen, it stays real, only on the condition that it continues to happen. (p. 75)

Wild foraging, in urban areas, provides ample opportunity to recognize that our relationship with plants and wildlife and soil, cannot be static. There can be no "finite" boundary to encapsulate our relationships. It is "without limit," as Berry explains. It is to be experienced by being present and attentive in the moment. As Poe and co-authors conclude in their study of forging in Seattle, it is a "communicative project not only between different groups of people, but also between people and more-than-human nature" (Poe et al. 2014, p. 915). Foraging also reminds us that food is more than "commodity" or "calories" (Williams 2016). In essence, foraging is about sensitizing our senses especially the visual, the olfactory, the touch, the auditory, and the taste. But first, what and how we choose to "see" makes a difference. If we pay attention with care, by walking a one-mile radius we can gather plenty of food and in the process, we can build community, eat healthy, enhance wildlife, and *become one with our place*. Foraging might actually support the development of a sense of place, by fostering a commitment to locale, as reciprocal relationships are nurtured with *life-bearing* entities. According to David Gruenewald (2003), places "teach us who, what, and where we are, as well as how we might live our lives" (p. 636). When we know our place, including the plants, the wildlife, the soil, and the human community, we tend to be more in tune with who and where we are. Contrary to common misunderstandings, with wild foraging in urban areas we are more likely to convey an ethic of care and less likely to act irresponsibly and wreak social or ecological havoc in our communities, as attachment can result in emotional bonds between people and places.

Being grounded in place refers to a reciprocal relationship in which one nurtures and is nurtured by the surrounding social and ecological environment. A tree provides a good example. As my co-author Jonathan Brown and I share, "While it is rooted in specific soil, bounded by contingencies of water, air, sun, space, and so forth, a tree at once contributes to shaping its own environment through shedding water, casting shade, and dropping leaf mulch. Thus the tree and its terrestrial home are intimately linked, each contributing to the life of the other" (Williams and Brown 2011, p. 58). As we experience while walking to forage, it is often through direct experience and investigation of the flora and fauna, the soils, the seasons, the rhythms of natural cycles, the histories, and the communities within which we humans live, that we develop and begin to feel a sense of place. For poet Gary Snyder (1990), "the small lessons, the enormous lessons, the lessons that may be crucial to the planet's persistence" are learned in interaction with place (p. 26). He urges that we intimately reacquaint ourselves with habitat. "To know the spirit of a place is to realize that you are a part of a part and that the whole is made of parts, each of which is whole. You start with the part you are whole in" writes Snyder (1990, p. 38). The logic of global capitalism instigates endless relocations and smooths out variations between places. Local context can be marginalized; conversely, interdependent links between local and global are increasingly evidenced by climate change, financial meltdown, and food contamination. Wild foraging in our urban area is a worthy place to "start with the part you are a whole in," as Snyder states.

I am not a member of any formal urban foraging communities. Until I was invited to write a chapter for this book, I had paid attention, only in passing, to the "trends" related to this phenomenon where I have lived since 1990: Portland, Oregon. In other words, being a gardener, my interest in gathering food in the wild started mostly with walks I took in the neighborhood and the discoveries my family and I made as we met large numbers of people and even larger numbers of trees and shrubs, vines and herbs, along with their wildlife cousins, on our walks. Perhaps the fact that I garden stirred an even deeper curiosity about and connection with the magnificence of what was growing "wild." Looking for food was not the motivating intent. Rather, the goal was to walk out of curiosity to get to know my watershed, my neighborhood, human and non-human, and to possibly establish roots in a new place and to belong. Foraging for food brought the not-consciously-sought connections. Dislocated from birthplace and culture, perhaps my soul longed to understand: Who am I? Where am I? Where do I belong? Plants and I speak a "language" like no other. Foraging made it an interactive adventure the lure of which has been irresistible. With each walking step, "the observer and the observed" have indeed changed. Walking a one-mile radius has offered never-ending possibilities to learn, to marvel, and to love. I am humbled as I revel in the complexities offered to me by each step and each encounter with the *life-bearing* and *life-giving* kin that enhance my embodied understanding of food and place.

References

Berry, W. (1991). *The unforeseen wilderness*. Berkeley: Counterpoint Press.

Bluehorse Skelton, J. (2008). *Green path to health and healing in urban Native America*. A Master's Research Project. Unpublished. Portland: Portland State University.

Cuba, L., & Hummon, D. M. (1993). A place to call home: Identification with dwelling, community, and region. *The Sociological Quarterly, 34*, 111–131.

Gruenewald, D. A. (2003). A multidisciplinary framework for place-conscious education. *American Educational Research Journal, 40*, 619–654.

Gros, F. (2014). *Philosophy of walking*. Translated by John Howe. New York: Verso.

Masumoto, D. (2003). *Four seasons in five senses: Things worth savoring*. New York: W.W. Norton.

McLain, R.J., Buttolph, L. P., Poe, M. R., Hebert, J., Patrick, T., Brody, Dzuna, M., Emery, M. R., Marla, R., & Charnley, S. (2012). *Gathering in the city: An annotated bibliography and review of the literature about human-plant interactions in urban ecosystems*. Portland: United States Department of Agriculture, Forest Service, Pacific Northwest Research Station General Technical Report PNW-GTR-849.

McLain, R. J., Hurley, P. T., Emery, M. R., & Poe, M. R. (2014). Gathering "wild" food in the city: Rethinking the role of foraging in urban ecosystem planning and management. *Local Environment, 19*, 220–240.

Oliver, M. (1993, August). *New and selected poems*. Boston: Beacon Press.

Orr, D. R. (1992). *Ecological literacy: Education and the transition towards a postmodern world*. New York: SUNY Press.

Poe, M. R., LeCompte, J., McLain, R., & Hurley, P. (2014). Urban foraging and the relational ecologies of belonging. *Social and Cultural Geography, 15*, 901–919.

Pollan, M. (2001). *Botany of desire: A plant's eye-view of the world*. New York: Random House.

Snyder, G. (1990). *The practice of the wild*. San Francisco: North Point.

Williams, D. R. (2016). *Beyond calories: Food as relationship*. Paper presented at the American Educational Research Conference. Washington, DC.

Williams, D. R., & Brown, J. D. (2011). *Learning gardens and sustainability education: Bringing life to schools and schools to life*. New York: Routledge. https://doi.org/10.4324/9780203156810.

A Sample of Sources and Google Sites on Field Guides and Wild Foraging:

Brill, S. (2012). *Wild edibles*. Mobile app for Android. New York: Winterroot LLC.

Brill, S., & Dean, E. (1994). *Identifying and harvesting edible and medicinal plants in wild (and not so wild) places*. New York: HarperCollins.

Craft, D. (2010). *Urban foraging: Finding and eating wild plants in the city*. Memphis: Serviceberry Press.

Lerner, R. (2013). *Dandelion hunter: Foraging the urban wilderness*. Guildford: Lyons Press.

Phillips, C., & Welty, E. (2013). *Fallen fruit, an interactive map*. http://www.npr.org/sections/thesalt/2013/04/23/178603623/want-to-forage-in-your-city-theres-a-map-for-that

Thayer, S. (2003). *The forager's harvest: A guide to identifying, harvesting, and preparing edible wild plants*. Omega: Forager's Harvest.

Vorderbruggen, M. (2016). *Foraging: Idiot's guide*. New York: Penguin Random House.

Dilafruz R. Williams 's passion for gardens is evident in her own delight and engagement with soil and life in its multitude of manifestations of wonders and mystery. Her writing has focused extensively on garden-based education – its conceptualization, articulation, and practical effectiveness. Her co-authored book, *Learning Gardens and Sustainability Education: Bringing Life to Schools and Schools to Life* (Routledge, 2011), presents curricular and pedagogical examples drawing upon her visits to over 100 education gardens across countries and continents. https://sites.google.com/pdx.edu/dilafruz/home

Chapter 3
In Pursuit of Elk

Joel B. Pontius

Encounter

The dark hair on his neck is dripping water. I can see this because I am standing in the grass right next to him, rocking with the rhythm of his antlers as he grazes. I am eleven years old and this is my first experience with an elk. "Joel, time to get back in the van. We have to go to the campsite – we're making fajitas for dinner," my mom calls from the rolled down window of the teal Chevy Astro. I want to spend more time with the bull. Slowly, I back towards the vehicle, dragged into the ordinary.

I watch the bull disappear out the back window as my dad drives around the parked vehicles and crowds of Rocky Mountain National Park. Back at our campsite, I split kindling to help my dad start the fire in the ring next to our pop-up camper. As the flames grow, my mom removes the plastic wrap from the Styrofoam dish of beef round steak, cubes the meat, seasons it, and pours it into the cast iron pan. It sizzles as Dad adds sliced onions and bell peppers. When the fajitas are done, my sister opens a bag of flour tortillas.

"Thank you, God, for our food, and for our day in the mountains," Dad prays as we hold hands in a circle around the fire.

"And thank you for the elk," I interject to close.

For many years I circle back to the time with the bull to soothe the boredom of sitting in a desk at school, or watching the Sunday school teacher stick Zacchaeus to the flannel graph in the church basement. Even when I search beneath rocks in the

Author's note: Thank you to Ann Hostetler for her wise perspective and editorial work on this chapter.

J. B. Pontius (✉)
Sustainability and Environmental Education Department, Goshen College, Goshen, IN, USA

creek close to home or hunt rabbits with Dad, I invoke these moments with the elk. I felt different when I was with him.

Immersion

At the Teton Science School's Graduate Program in Place-based Education in Grand Teton National Park, Wyoming, I wake early most mornings, grab my new binoculars, and clamber down the stairs from my room above the maintenance shop to walk and look for animals before the air warms and the day's programming begins. These slow walks show me where moose and mule deer browse bitterbrush, how western meadowlarks bundle the sound of sagebrush, and when the small herds of elk – who stay mostly hidden in the folds of terrain – negotiate the edges. Eleven years after my encounter with the bull, I am living in the Greater Yellowstone Ecosystem – home to the largest elk herd in the world.

In one of my first formal experiences as a place-based educator, I sit at the edge of an expansive meadow with a dozen fifth graders. In the hazy September light, we watch a cow elk and her calf emerge from the forest at the base of the Tetons at sunset. *Heeaaww* – the long-nosed cow calls to the elk behind her in the cooling air. Elk after elk follow her line and spread out into the meadow. The last animal to step out from the trees is a mature bull whose antlers are blackened with pinesap except for the white tips, polished by friction. From the edge of the trees, he tips his head back, opens his mouth and bugles a high-pitched squeal before herding the cows and calves into a tight clump.

While the students tally on paper the elk vocalizations they hear, I am focused on the elk. I have been waiting years to spend this time with them. I am here to learn in this community of naturalists, and to immerse myself in the daily lives of elk.

Hunting Mentors

I grew up hunting. It is a practice tied to many relationships and places, politics, spirituality, identity, and other facets of my life. When I am twelve, I begin hunting white-tailed deer with Dad around home in Bloomington, Indiana. The morning of my first hunt – a rite of passage – Dad is out of town, but he arranges for one of his friends to take me. It is snowing large flakes as we walk through the woods in the darkness. When we reach the red oak tree that the stand is hanging in, my dad's friend screws in the three bottom tree steps and ties a rope to the strap of the shotgun I am carrying. The steps and the stand are slick with snow and ice, and I hold the rope tightly as I climb step by step to the two-square-foot platform that is about 20 feet off the ground. I pull the gun up to the stand and my dad's friend whispers that he'll come back to get me when it's time to go.

I sit in the dark, excited and afraid, waiting. As the sun rises, a young buck walks up the hill from behind me. I aim, shoot the deer, and watch him run over a rise and out of sight. I climb down from the stand and I am stunned when I find large droplets of blood where the buck stood. The bleeding grows heavier as I follow, and after a short trail I see him lying dead on the edge of a ravine. I am devastated and relieved at the same time. I kneel down and put my hand on his side. As my tears start to flow, my dad's friend shuffles in. He had heard the shot and came to see what happened. He sighs and moves quickly to the dead deer, lifts his head by one antler and drops it to the ground. "You should've let him go. His antlers aren't big enough," he says sharply. I guess I did the wrong thing. I thought this was about the deer meat. I concentrate on holding back the tears that should have been shed while we drag the deer's body out of the forest.

My dad comes home later that afternoon and he tells me how proud he is. "Great job, Joel – and you were only in the stand half an hour before the deer walked by?" Dad and I drop the animal off at a butcher shop. There are hundreds of deer laying out on a parking lot with hunters' names on them, and just as many hanging by their hind legs from the rafters of a metal-sided pole barn. Men are walking around touching the largest deer antlers. A month later, we have a message on the answering machine from the butcher, "Deer's done. Come get it. It'll be $75 cash." The butcher rolls a cart with three stuffed paper grocery bags from the walk-in freezer. I help Dad make steaks for dinner that night, proud to share the meat with my family. Dad buys an instructional video on butchering that winter, and from this point on, we work out our own butchering process in our neighborhood backyard, hanging deer from a branch in the blue spruce.

At fifteen, Dad buys me a bow. I spend time in the backyard every evening learning to use it. I have never enjoyed guns, though they can be useful tools for hunting. I like shooting my bow because of the focus and rhythm. Bow season is six weeks long and makes many early mornings and late evenings of quiet in the woods. As a teenaged boy in the Midwest, I need this time to slow down and watch the world move. One of my first evenings bow hunting, an orange-haired doe walks a creek edge close to me. I observe her careful selection of plants – she uses the side of her mouth to eat them. I'm captivated by watching her and forget about my bow. Dad comes to pick me up from the stand and sees the doe standing next to me before she runs for cover.

"Why didn't you shoot?" he asks when I reach the ground.

"There were branches in the way," I lie to him.

I struggle to express my competing desires to be close to animals and to hunt them. The hunters I know talk about how important conservation is, but it does not seem to me like they are interested in much beyond the bones that grow from a deer's skull.

Kevin

"What is that made from?" I ask, motioning at the tether from which Kevin's sunglasses dangle. Kevin is my boss at Wildlife Expeditions, the ecotourism group I am guiding for in Jackson, Wyoming after my year at the Teton Science School. His hair is long and matted, not quite in dreads, and he wears a pair of weathered, sage green Carharts.

"Interesting that you would notice that," he says with a measure of intensity and a wide grin. "It's rawhide from an elk I hunted. It's part of how we eat. I want to make sure the animals know how grateful I am, so I use everything I can."

Kevin uses elk hide like duct tape to fix tool handles, make journal covers, and as cordage. I resonate with his reverence for the animals, and that he hunts them for food. He is a forager, too, and grows as much food as possible in a garden. It is his goal to eat something that he hunted, foraged, or grew at every meal of the year. I am inspired by this. Every time I ask Kevin a question about his practices of hunting and foraging, he answers with a story that highlights the intelligence of the land. He has found me at the moment in my life when I need his teaching.

Wolves

Through binoculars, I watch a herd of fifty elk graze in a meadow below granite peaks in mid-July. On the edge of the group, two calves kick up dust as they chase each other. From where I'm standing, I see five wolves loping toward the herd. As the pack breaks the rise, the herd bolts. In mere seconds, a young bull is separated from the herd and a wolf hangs by its teeth from each of his four legs. As the fifth wolf reaches the elk, it latches onto the elk's throat. Gradually, the bull collapses under the gripping weight. Above the sound of the wind and pulsing tires on a distant road, the wolves howl, each at a different pitch – mellow and round.

In this moment, I feel hungry as my gut senses that everyone is about to eat – not just the wolves. Eagles, ravens, magpies and coyotes will find what is left of this bull in the morning. After the kill, the elk herd turns back to filling their rumens – the bacteria-rich holding chambers of their digestive systems – with grasses and forbs that they will digest when they bed.

If I am living here in the Greater Yellowstone Ecosystem, I want to be part of this exchange. I spend the rest of the summer preparing for my first elk-hunting season, maps spread across the floor of my apartment. Going mostly on intuition, I consider the structure of the places elk might move through, why, and when. I check my map work when I am out in the valley guiding wildlife viewing trips. As I look ahead to my first hunts, I overhear a man and woman who appear to be in their late 20s in the fly shop. They talk about many elk seasons without success, about a difficult process that's nonetheless worth pursuing. The hunting cultures in this place are different from the one I experienced growing up. Elk meat is prized as exceptional food, hunting often means extended trips into the backcountry, and does not necessarily involve wearing camouflage. For some Jackson residents, hunting elk is an extension of living alongside them.

Learning to Hunt Elk

On my first bowhunt for elk in the Absaroka Range, I am up most of the night listening to the elk bugle. One bull in particular has bugled on and off for hours. Well before first light, I burrow down into my sleeping bag with my headlamp and map to think through how I might approach him. It will depend on which direction the wind is blowing from – I have noticed that the wind direction changes often during sunrise. An hour later I walk upstream over river rock in a dry streambed toward the valley where I hope to find the bull.

When I feel like am close, I take my bow in my left hand, reach for a stone with my right, and carry it up the embankment. To sound like a herd of elk moving through, I roll the stone back down the embankment – *fup, fup, crack* – and cow-call from the latex reed I am carrying in my teeth. I see pine boughs moving on the other side of the flat, and a flash of antler as he runs straight toward me with his chin to the sky. I am about to dive out of the way when he stops suddenly on the other side of the pine tree where I am standing. Overwhelmed by his presence, I forget to draw my bow. He ambles into the streambed looking for the group of elk he thought he heard. My legs are shaking from the rush of energy that the bull left in his wake. I sit down to breathe and steep.

Although I had internalized many of the elk's patterns of movement and social habits through a year of observation, only gradually did I realize that I could communicate with them in ways that they responded to. For the next several weeks, when I am not guiding wildlife-viewing trips, I lose myself hunting elk.

First Elk

It is still cold in the trees at noon as I follow a worn game trail through the steep north-facing forest near the peak of a mountain in the Gros Ventre range. I have climbed since 4 A.M. to get to this place – a winding ascent of three thousand, five hundred vertical feet. I sense that I might be close to elk, and also that I should turn back soon – I fear I have wandered too far. In the next moment, I see fresh elk scat in the center of the path, and smell the musky, yarrow-like scent of elk in the air. I kneel down and cow call softly.

My first glimpses of the animal are his dark legs gliding through the timber. He walks quickly into an opening in the trees, and pauses to look for the cow. Before I fully realize it, I have taken my shot and the bull dies in the pine duff. I sit with him and run my hand over his mane. I talk to him for some time as if he can hear me.

I don't know how I will make it off this mountain with him. It occurs to me that I am so high up on the north face that I might have cell phone reception. I do. My partner, Laura, answers, gathers some friends to help, and heads my way.

I am able to skin and quarter one side of the elk before I'm exhausted and out of water. Five hours later, I hear voices on the ridge and climb up to meet my friends. "It's way farther up here than it looks from the trailhead," Laura says as she gives me a hug. "I was just following my gut. Do you have any water?" I ask. She hands me the blue bite-valve of her camelback and I take several gulps. On the way down to the bull, we scoop snow into what is left of the water to stretch our supply.

We reach the bull and stand around his body in silence for a while. Then, they help me turn the horse-sized animal over to skin and remove the meat from his other side. We need headlamps by the time the bull's legs, tenderloins, other edible meat, hide, and antlers are loaded into or strapped onto backpacks. We have to pack out through the forest. The south side of the mountain is open, but steep and with scree slopes that would be dangerous with weight on our backs. My pack weighs over one hundred pounds. It would have been wise to stash the meat in the trees and come back for it in the light, but I was afraid that a bear or other animals would take it, and we were walking back to the trailhead either way. I struggle to keep moving under the weight, and as I crawl over fallen trees I need help standing up. Through the night, the stars start to waver and the breaks we take become longer. We reach the trailhead in the early morning.

The next afternoon in the single car garage in our apartment complex, I work the blade of my boning knife along the edges of the heavy muscle groups in one of the hind legs, releasing them from each other. I am in awe of the color and consistency of the muscles and their size. We don't have much in the way of processing supplies, so I place thick sirloin tip steaks in our kitchen pans, and certain roasts in certain pots or on cookie sheets. As I finish wrapping the first quarter's meat in freezer paper, I consider what to do with a bowl full of trimmings. That night, we have fajitas.

I call Kevin to tell him the news. "You dog! Congrats! Well, butchering the elk is some of the most important work you'll do all year and you have a responsibility to the animal to do it well. Concentrate on that for the next few days – we have everything covered here at work," he offers. I need the time and still carry the lesson.

The highs are in the low 30's the next three days, so I leave the meat in the garage and butcher slowly. During the process, I buy a small block freezer for $99 at the Sears in downtown Jackson. It is filled completely by the bull, and we have to store the overflow elk in our kitchen freezer. During the fall and winter, elk meat is part of all of our celebrations, and we eat small portions maybe twice a week. As we eat, we retell the story and revisit the place and process in our minds.

The next July, ribbons of arrowleaf balsamroot wrap the ridgeline as Laura and I slowly climb the five miles and three thousand, five hundred feet of elevation to revisit the place where the bull died in October. The walk is humbling – the paradox in all of this. In daylight, we piece together the struggle to carry his muscles and hide from the mountain. We find most of the bull's spinal column and ribcage about 100 yards downhill from where we left him. Even the cartilage on his ribs has been eaten, leaving blunted bone ends. The undigested plants that were in his rumen when he died are spread in a five-foot radius, and there is a pile of scat on the edge, with bits of seed and hair. What a meal this bear had sniffed out among the pines.

This is our last hike as we bring our time in the Greater Yellowstone to a close and move to Laramie along Laura's path toward a law degree and mine toward a doctorate. Elk herds have become an important part of my life. I have read that there are strong populations in southeast Wyoming. As we move to a new place, they will be part of my orientation to the hardscrabble mountains and shortgrass steppe.

Tracking Deeper

Listening

Light seeps in from the east and the clouds crack open, pouring snow on the feet of the North Laramie Mountains. I walk into the aspen-covered valley with an heirloom rifle in my arms and angle up the mountain. A strong whiff of elk scent draws me into a line of tracks that cannot be more than a few minutes old. One line leads into more, then the tracks ribbon over the ridgeline. I start to follow when beyond the veil of falling snow, I hear: *kraaaaa, kraaaaa* – the throaty voice of a clark's nutcracker. An image of the bird roosted over a herd of elk flashes to mind. *Kraaaaa, kraaaaa.* The image, again. The tracks look like they are going a different direction than the bird is calling from, but I sense that I need to listen to the bird.

I am between a sheer cliff face and piles of granite boulders that had fallen from it when the snowfall slows, silhouetting an elk across the ravine. I lower myself down to the rock, becoming aware of the elk herd, bedded with snow-covered backs. A bull and a cow graze on the dried yellow grasses poking out of the snow. I have a population reduction license for a cow. She falls with the shot and dies in the snow while the herd moves up the ravine and over the mountain.

Hours later, I don my pack full of meat and a clark's nutcracker lands in the tree next to the elk. Before I walk out of sight of the elk's body, I stop and look back. The bird's gray head bounces rhythmically on the elk's ribcage. As I plod through the snowy valley, ravens fly overhead to join the feast. A few years before this experience, I would have had serious reservations if a person had told me a story about birds leading them to elk herds. I am slowly learning that the landscape still addresses humans as animals, whether we hear this or not.

Responding

The winter herd of more than one hundred elk moves up the slope in long lines, over two miles away. I slog through snow that is up to my knees in places hoping to catch up with them and taste blood in my breath from overexertion in the fifteen degrees below zero air. I'm moving slowly by the time I reach the ridge when I see a group of ravens fly over the rise. They dive and twist in a playful way I had seen before in the Lamar Valley of Yellowstone when the ravens danced to convince a pack of wolves to hunt elk.

I am almost sure that the ravens are showing me where to find the herd of elk. They circle for a minute, spiraling within 20 feet of the top of my head before they fly over the next rise. I follow them, sneaking through a field of boulders into a place where I can peer into a loosely growing grove of aspen trees. The last ten elk in the herd browse through. The ravens have timed everything just so.

How often do ravens, or other species, address us? What are we losing as a species as we distance ourselves from our own abilities to join these conversations?

A few days later, I am looking out the west-facing garage window while I debone the elk's hindquarter on my butchering table. As I remove muscle groups, I think about the cuts I will make for particular recipes. From the south, a raven flaps down and lands on the elk scapula I had laid out for them in the snowy yard. Another flies down from the light post across the street, snatches a ribbon of sinew, and caches it under the base of a low-growing juniper. I imagine how the area appears to them from the sky – in this straight line of duplexes, the arching elk bones and the color of muscle. I watch the birds feed while I work.

Shadow

Laura's body made our daughter's, so mine should contribute something. I learn from a friend how to build a crib for her out of dead-standing aspen gathered from the Snowy Mountains west of Laramie. I had just sloshed through a subalpine bog when I stumble upon the front right post of the crib – a tree that has rows of teeth marks dragged into it from an elk's meal late some winter. After I peel the bark with a drawknife in my garage, I arrange the bug trails and wood grain so that our child will be held by these forms as she falls asleep, while she sleeps, and when she wakes.

With the crib complete, I go to the mountains for the opening day of bow season for elk, hoping to bring home food for my partner and new child. There are lots of other priorities to attend to this September with our child coming into the world, but I walk miles into the mountains, to a place I know at the confluence of two streams, and listen to elk bugling in the dusk while I set up my tarp. I wake to elk movement at 4A.M., so I make some oatmeal and coffee. As the water heats over the blue flame of my backpacking stove, I gather grasshoppers slowed by cold, and mix their protein-rich abdomens into the oats. A bull bugles in the darkness across the canyon, and is answered by other bulls from three directions. I shove a day's worth of food into my pack, filter water from the stream, and make my way to a place where I will be able to see the surrounding mountainsides.

In the first moments of light, I see the pale rumps of a herd of elk grazing along a steep mountaintop. A sliver of forest funnels from the bottom of the slope to their location. Ducking under pine boughs as I hurry through the drainage, I keep a mental image of where I saw the herd. This is challenging to do, as the perspective changes with every turn. To reach the elk, I have to pass through an open meadow. There is no way around the herd seeing me, but if I hide my silhouette they might let me pass. I don't have another option. I crawl out into the meadow and the elk stop and look down at me. We are about one quarter of a mile apart. I know they cannot smell me from this distance, so I keep crawling along. The herd allows it and continues feeding.

In the cover of the trees, I climb as quickly as I can. When I am close, I have to stop to catch my breath. For the last one hundred yards, I can see three elk grazing, and I have to wait for the right moment to make each move from tree to rock to tree. The closest form to the elk is a charred limber pine snag that looks like it has been burned by a lightning strike. I crawl the last fifty yards and kneel behind the snag. Three cows graze among the uphill boulders, one noticeably younger than the others. She will be the best for food, and less important to the herd than the older cows. I wait for an opening. She steps out from behind a granite rim and lowers her head to graze. I draw my bow, aim, release, and watch the arrow's arching flight. As it strikes the elk, she jumps and trots downhill toward me. The cow beds within steps of me, and lies her head down on the soil. The rest of the elk continue grazing.

I feel like I am finally starting to understand these animals. I place my bow on the ground, and swing my pack around to have a sip of water. The wounded cow, who I have taken for dead, sits up and looks directly at me. We hold eye contact. I do not know how she finds the energy to jump to her feet. She runs wounded into the shadows.

An oval pool saturates the place where she bled into the ground. I follow blood droplets for a quarter mile until I can no longer find them. Hundreds of elk tracks are strung in every direction in the dust on the hillside – I cannot tell which tracks are hers. I walk concentric circles for hours, coming back to the last droplets of blood, and find no more. So, I begin wandering to places that I think she might have gone. In the process of searching for the rest of the day, I unravel. Somehow, I manage to lose my GPS unit, my fallback in case I get lost. Then, I realize I have left my map and compass at the tarp shelter. I sit on the side of the mountain in a place that I thought I knew, lost, ashamed, and worried that my mistake had ended the cow's one life.

After resting for a while, I drag myself higher to look for anything familiar. I see the burnt limber pine snag miles away, and I am reoriented just before last light. I was dangerously off track. The next afternoon I make it to my vehicle, and lift my pack onto the coolers full of block ice that would not cool elk meat. It is possible that she survived, but she likely fed someone else's family. What does it mean that this is the story I have to tell mine?

Fig. 3.1 Recipe: Emerson
June's baby food

Recipe: Emerson June's Baby Food

2 pounds of elk front shoulder roast, cooked 6-
8 hours on low in a crockpot
4 cups peaches, apricots, or cherries, sauced
Other dried fruits can be added
Cinnamon to taste

Puree cooked elk meat and sauced fruit
together, roughly ½ meat and ½ fruit
Pour into ice cube trays and freeze
Thaw to serve
Thin with breast milk or water

Baby Food

Soft light illuminates my 9-month-old daughter's hair against the midsummer aspen leaves as I lift her into the green carrying pack. She is happy when we're out together, and rides along quietly for hours of trail time. After a mile or so of track-ing, I find fresh elk scat paddies. We are close. Emmy reaches out for a fir branch and feels the needles. The distant crack of a breaking stick signals the herd's loca-tion. We move to the next tree, and the next. The elk scent thickens before I see their thin, dark, summer-maned necks moving through the tall grasses.

We draw in close to the base of a subalpine fir to make sure we are not silhou-etted. Thirty elk or so – cows and calves with young bulls intermixed – make up this herd. There are too many eyes for us to move closer without spooking them. It does not matter. The elk move toward us. When Emmy sees them, she kicks her legs and scratches her fingers on my shoulder in response. For half an hour the herd sur-rounds us. Then they flow south (Fig. 3.1).

I remove a thermal bottle of hot water from my pack, and two pyrex dishes – one larger than the other. In the smaller dish are three cubes of baby food, a mix of apri-cots and elk meat pureed and frozen in ice cube trays. I pour the hot water into the larger dish, nesting the smaller one inside to defrost her food. We sit together on the granite and eat.

Last Elk

Elk scent drifts on the wind. I kneel down in the snow near tree line in the Snowy Mountains west of Laramie, and nock an arrow. The snowstorm is becoming a blizzard and the elk are moving lower on the mountain in response. I am here to connect with this movement. I see a cow elk, then the tips of her calf's ears bouncing beside her. They pause next to me as they eat together. I watch intently, but do not draw my bow.

I cannot separate this elk mother and child. They walk upslope to the north, and when they are well out of view, I follow their tracks across a streambed that is drifting over with snow.

I wait for the cow and calf to move on and then follow their tracks that lead into a grove of spruces. Along a tangle of windfall, I see what appear to be elk antlers sweeping back at a strange angle. I move closer. An old bull elk lay on the ground, dead, with his nose to the sky. As I stand there wondering how he died, the bull lifts his head and slowly struggles to his feet. His eyes are sunken and he is so emaciated that I can see the tops of his hipbones. Whether it is the right decision, I do not know, but I cannot walk away. I nock an arrow, draw my bow, aim for his heart and release. *Whump*. He collapses – I hear his last breath. Searching his body for signs of what had happened to him, I find some matted hair on the side of his belly. When I touch it, puss seeps from a wound. I look closer and realize that he had been maimed by an arrow. It must have happened weeks earlier for the bull to lose so much weight.

As the storm swells, I salvage the meat that I can. Skinning the bull, I nearly vomit due to the smell of infection and have to take short walks away from him to work out the nausea. I struggle to stay on track the many miles off the mountain in the storm. Even with all my layers on and carrying a fully loaded pack, it is difficult to keep warm. Back home, I am not surprised that the meat tastes awful. I eat it anyway. Through this experience, I understand with more subtlety what may have happened with the cow elk I wounded years earlier. I see both of their faces as I chew.

This wounded bull is the last elk I bring home to eat. After finishing my doctorate, I move with my family back to Indiana to be closer to our extended family. What I learned from the elk in Wyoming, I brought with me to look anew at a landscape I had known since childhood, where I would teach others to listen to the more-than-human world while being part of it.

Ancestral Memory

When the cottonwood leaves around home in Northern Indiana begin to yellow, I receive a package from my mom. My daughters pull the packing tape from the box, and we remove an old leather case. I run my fingertips along the hand-stitched edges, then open the lid to find a pair of binoculars. Inside the lid, lined with faded blue velvet, I see stamped in gold foil:

Joh. Conr. Schmidt
   Optiker
   Nürnberg

Mom has slipped a note into the box:

"Dear Joel, Your Grandpa thought you might like to have these binoculars. They belonged to your great, great grandfather Frederick Carl Verch. He was the first in my bloodline to migrate from Germany to North America, and these binoculars were one of the few possessions he carried with him on the ship. Grandpa said that Frederick carried them in the foothills of the German Alps as he hunted for red deer each autumn. It made him think of you and the elk."

Elk are a highly evolved species of red deer that adapted to grazing in Siberia and crossed into North America at the end of the last ice age; they are very closely related to each other. My great, great grandfather hunted red deer for part of his life and diet, and I imagine other ancestors of mine did as well. Perhaps ancestral memory is part of why I connected so strongly with elk as an eleven-year-old, and later followed them along a deeper path into place, land, and food.

I understand why my great, great grandfather chose to bring his binoculars with him. The associations binoculars carry with them – the places, the animals, the ways the terrain comes together from a distance – continue to orient even in unfamiliar landscapes. I wonder where he walked with them, what he thought about, and if he, too, felt a kinship with the land and animals he saw through them. Of all the movements and relationships that we embody, I wonder how many come to us by way of our ancestors.

Encounter

"He has beautiful antlers, Papa!" seven-year-old Emmy gasps as she looks through my binoculars. We are at the base of the Tetons, visiting as a family, and I have brought her to one of my favorite places to watch elk herds. Wide-eyed, she continues, "There's something special about elk."

I nod. "Did you know that elk are part of your body? Elk was the first meat you ever ate," I offer. Squinting slightly, she concentrates on watching the bull for several more seconds, lowers the binoculars, and looks at me.

"Is that why my legs are so strong?"

"Probably," I respond.

"Can we stay here and watch him for awhile?"

"For as long as you'd like."

Joel B. Pontius is a sustainability and environmental education professor at Goshen College's Merry Lea Environmental Learning Center. His writing, teaching, and creative work live at the edges of spirituality, sustainability, and the more-than-human world. You can contact Joel at joelbpontius@gmail.com; Instagram@joelpontius; or on his website: www.joelpontius.com.

Chapter 4
Thanksgiving in Alabama: Deer Hunting Among Paradoxes in the Black Prairie

Abbie Gascho Landis and Andrew Gascho Landis

"Tell me about the first time you went hunting in Alabama, Andrew."

That first Thanksgiving I spent in Alabama, I was fifteen and had never fired a gun before. Growing up in western Pennsylvania, we had close family friends who had five kids, including two boys the same age as my brother and I. They spent every Thanksgiving with their granny and granddad in Alabama, splitting time between the big house in Tuscaloosa and sprawling hunting properties an hour away, in Greene County. In 1993, they invited me to join them.

My dad never hunted, but I grew up in a culture that did. We always had the first day of hunting season off school, and my classmates returned from hunting with stories of deer they saw. If they didn't have stories, they invented some. It seemed natural for me to be interested in hunting since I loved being in the woods. When our friends invited me to Greene County, I wanted to go. My dad talked with me before I went. We were standing by the coal bin at the school where I was in ninth grade, and he was the principal. "Do you think you could actually shoot a beautiful animal like a deer?" he asked me.

"Yes, I think I could," I told him.

The invitation to hunt in Alabama appealed to me in ways I only later began to identify. As a kid, I relished outdoor adventures, losing myself for hours in the woods by our house or camping with my friends or family. I devoured books about wilderness survival and homesteading skills. Hunting deer for meat seemed to promise gleaning food with adventures in the woods. I felt both excited and a bit nervous to have a chance to prove myself in this way. Also surrounding the trip was my anticipation of visiting my friends' Southern culture, including their favorite Dreamland Bar-B-Que and The Cotton Patch restaurant.

In Greene County, my friends' granddad owned four properties totaling around 5000 acres, and over the next twenty years this patchwork land would become important to me in ways I could not have predicted. They called their original

A. Gascho Landis · A. Gascho Landis (✉)
Fisheries, Wildlife & Environmental Science, The State University of New York,
Cobleskill, Cobleskill, NY, USA
e-mail: gaschoam@cobleskill.edu

© Springer Nature Switzerland AG 2020
J. B. Pontius et al. (eds.), *Place-based Learning for the Plate*, Environmental
Discourses in Science Education 6,
https://doi.org/10.1007/978-3-030-42814-3_4

property the Old Farm. It was a scrappy 750 acres, with several large, brackish cat-fish ponds, cedar scrub, and just a few spots that supported fast-growing pines. The largest property was 3200 acres called the Matthiessen Place, which had once been plantations—soybeans, grains, cotton—with fertile soil thanks to a creek running through it. It had a beat-up history and a varied landscape, mostly black prairie. On the 1100-acre Willis Tract, hills bordered the old cropland, and there were hard-wood trees and a set of abandoned homes across one knoll (Fig. 4.1).

On opening day of my first year, the granddad, a retired doctor, invited me hunt-ing with him to the fourth property, called Mauvilla. The name, which is also the origin of Mobile, had been heisted long ago—as was the land—from Native Americans. This Mauvilla property—850 acres along the Black Warrior River—was not prairie; it was bottomland hardwood forest. I fell in love with Mauvilla and the way it transported me from prairie and scrub into ancient forest. One time, when just the granddad and I were there, a big fog rolled into the Mauvilla swamp. He couldn't see to drive, so I drove him home to the Old Farm with the stick shift. At seventeen, I remember feeling proud and useful, relied upon, competent in this less-tamed landscape of dirt roads and deer hunting.

"What gun did you have for hunting?"

"It was their .30-06, an old World War II, bolt-action rifle. It was a family favorite, passed around, since it shot nicely even though it was old. Someone had dropped it out of a tree stand once, so the wooden stock had snapped off, and they'd fiberglassed it back on."

Fig. 4.1 The Mattheissen Place

I shot it twice for practice and then the granddad set me by myself at a stand known as the Little Oak Patch. A large, enclosed deer stand sat at the bend of an L-shaped field, planted with tender green grass. The stand nestled against pines—also planted—and looked across the field to an older stand of mixed oaks and sweetgum trees. Back then, the Black Warrior River basin had tons of deer, and you could see forty or fifty at a time at Mauvilla.

"Be careful," the grandfather said, "Make sure you shoot a deer with at least six points." As sport hunters, my friends' family focused on curating a population of growing bucks more than gathering meat. He told me to pick a specific landmark where the deer left the field, if it ran (Fig. 4.2).

"So you sat out there by yourself and killed a deer?"
"I shot it."
"So then it was dead?"
"Well, it was mostly dead."

When they came back to get me, it was dark. We began tracking at my landmark at the edge of the field. As a light fog gathered, the woods smelled of decaying leaves. It seemed an impossible search, and that the deer was long gone, but I was first to find the deep red blood on some oak leaves thirty feet into the woods. Filling with relief, I lifted my flashlight from where I stood to look deeper among the trees, and there he was, lying down, staring at me.

I approached the deer and held the gun point-blank to end his life, but couldn't see anything through the sight in the dark, so I felt confused that the deer still sat upright after I pulled the trigger. Somehow I had missed him, even at this close

Fig. 4.2 Typical deer stand

range. The granddad took the gun from me and held it like a pistol, and didn't miss. At the loud, ringing shot through the neck, the deer slumped. The granddad congratulated me quietly on my first deer, and we turned to the practical matter of needing to butcher the deer before the meat spoiled in the mild Alabama night. We threw him on the back of the blue four-wheeler, took it back to the Old Farm, and cleaned the carcass that night.

My friends' father was a first-career butcher, second-career medical doctor, who prided himself on cleaning and butchering a deer in twenty minutes. Later, I realized he could do this because he was never the one to cook the meat, and neither was I. He taught us to bag a whole shoulder and freeze it—a daunting thing for my mom to thaw and prepare for dinner. The granddad used butcher paper and neatly wrapped and labeled every piece of meat. Although they had an industrial meat grinder, we never used it in the first five years I hunted with them. Instead, I brought home plenty of big roasts.

Even though I only spent eight days of the year hunting in Alabama, we ate the venison all year round. My actions during that time seemed to echo into the rest of my life—on our table, with my friends, in my time in the woods of my Pennsylvania home. Although I had proven myself a good shot in Alabama, I tried hunting—without success—in the Allegheny Mountains close to home. Hunting on foot through thick forest was a wholly different experience than hunting from a deer stand beside a field of cultivated deer forage in Greene County. It was so different from my home—in ecology and in the hunting experience—the black prairie in Alabama captivated me and became part of my identity, just as its deer became part of my body.

Meat

> *"Abbie, your first year in Alabama was that year when* everybody *was there, right?"*
> *"Yeah. It was the free-for-all year."*
> *"Were you mostly horrified or entertained?"*
> *"Mostly both. And other things too. I was surprised by how much I loved all the people, how natural it felt to butcher a deer, and how subtle and muddy the landscape seemed."*

That first Thanksgiving I went to Greene County, I was twenty-two and had never fired a gun before. I had never held a gun. Guns repulsed me with their sinister ability to kill, as if they would go off unpredictably and fire bullets in all directions. The first—and possibly only—time I saw my dad with a gun, I was about six, and he stood on the long sidewalk leading to our farmhouse front door. Facing our gravel driveway and the field beyond, Dad lifted the gun to sight in a fox, ready to eliminate the Rabies threat like Atticus Finch in *To Kill A Mockingbird*. I remember tugging on his arm, begging him to spare a life, horrified by the sudden appearance of this loud, cruel weapon. *Charlotte's Web* heavily influenced me, and my heart did not have any acceptable place for aiming guns at animals. I easily discarded respect for boys or men who enjoyed hunting animals for sport, based on their obvious lack of compassion and integrity.

We did not eat venison at home, and I cherished disgust for meat from wild animals such as deer, rabbits, or squirrels, although I ate chicken, beef, and pork with gusto. One night, a family friend laughed triumphantly after serving me tacos with ground venison, passing it off as beef. My fury at his betrayal further tainted the idea of eating venison, and reinforced my image of hunters as unscrupulous people.

That first Thanksgiving that I flew from my vegetarian lifestyle in Tucson to spend the week hunting in Alabama, Andrew met me at the gate in the Birmingham airport. We were two, devoted years into dating. He wore an olive-green, floppy-brimmed hat, which he'd been wearing in the first photo my parents ever saw of him, posing with a soft smile and a rifle. The floppy hat would last him for the next few years, into our marriage, when our dog would eat it.

We drove from Birmingham down into Greene County, deep in the part of Alabama known as the Black Belt. Crossing this landscape that was clearly so embedded in Andrew's heart, I felt confused. It seemed scruffy and flat—not my type, and not matching my expectations based on Andrew's devotion to it.

His friends and their friends joined us at the Old Farm, where we all slept in a bunkhouse that leaned to the west. They showed me how to safely hold a rifle, set its hard butt into my soft shoulder, and to hold my breath, relaxing into stillness when I squeezed the trigger. Each time I shot at the target, the noise and the kickback into my shoulder surprised me.

Later, sitting with Andrew in a deer stand, I took aim at a doe. Where we sat was typical of the deer stands on this property, with a wooden staircase, not a ladder, enclosed with a patch of carpet on the floor, and some folding chairs. I leveled the rifle through the slot. Winking into the scope, I could see the doe's features magnified—the hair in her ears, the way her shoulder moved along her chest. I waited, holding my breath. I could not shoot. Andrew took over and fired efficiently, dropping the doe. Then a fawn came out of the undergrowth at the edge of the green field. I wept.

After the fawn faded into the waning evening light, we climbed down to get the doe, dropped lifelessly onto the green winter wheat. Back at the cleaning shed, I learned to hang her upside-down and start at the groin—tracing around the full mammary glands, slicing downward through the skin—then peel off the hide. We opened the abdomen, careful not to puncture the intestines. I learned to disarticulate, to follow lines of connective tissue with a blade, to find the tenderloins. From the whole process, I learned what it really means to eat meat, and I ate it. Although I returned to vegetarianism when I left Alabama, 'carnivory' seemed both delicious and appropriate that week, given my participation in the animal's transformation into meat.

As it happened that year, the hunting and butchering crew was a crowd, with an age range of about seventeen to twenty-three, predominantly male. The resulting dynamics—not surprisingly—included significantly more goofing around than other years before and since. To my relief, everyone was calmly serious when it came to handling guns, but work in the cleaning shed ranged from jovial to raucous. There was a deer transported from field to cleaning shed on the hood of a Toyota Tercel station wagon, leaving a predictable bloody mess. There was, at one point, a pouch made from a deer scrotum.

The whole experience was muddy and animal—a stark contrast from growing garbanzo beans in a drip-irrigated garden in the Sonoran desert. Strangely, I relished the messy intimacy with the landscape and the deer that became food, even while I struggled to digest it all.

Land

"I'm not sure I understood right away why that place was so special to you, Andrew. It's not where you lived growing up."
    "It's one of my favorite places, ever. The bottomland hardwood forests seemed like something otherworldly. Big trees, saw palmettos. It matched my idea of wildness."

My friends' five thousand acres sat halfway down the western side of Alabama. Running diagonally across Alabama—not far north of my friends' property—is a ledge called the fall line. It delineates the slightly higher elevation piedmont from the lower coastal plain. One crescent of coastal plain is known as the Black Belt, named for its thin layer of dark, fertile soil, and for the human racial demographics during and since slavery. The landscape is mostly flat, with some gentle undulations. Rivers run through it—the Black Warrior, the Tombigbee—and they have broad floodplains, swamps and wetlands level with the rivers themselves. In these bottomlands grow old hardwoods that add to the sense of richness and mystery (Fig. 4.3).

At the heart of the Black Belt, Greene County is the least developed county in all of Alabama. Its county seat, Eutaw, hosts the annual Black Belt Roots Festival, where musicians perform blues and gospel music alongside homemade quilts, crafts, and food.

I grew an attachment to this place in the same way that one week every summer on a grandparents farm can stick with a city kid. Early on, the granddad took us squirrel hunting down in Mauvilla. He wanted to get his grandkids out there, so we went in the evening, and he sat us all along a two-track dirt road. As we sat there among the huge oak trees, squirrels ran everywhere. Mauvilla was right next to the Dollarhide swamp, a 5,000-acre private hunting club. One summer, I drove in the backside of Dollarhide and found myself surrounded by huge, ancient trees, growing along the Black Warrior River.

It wasn't all hardwood forest on my friends' properties, and the contrasts among the pieces of landscape added to the beauty. There were cultivated landscapes, pines—plantations for later harvest—and greenfields—sown to draw deer into open areas under their deer stands. Even better, there were swamps too wet for hardwoods, and there was prairie.

Later, in college in Northern Indiana, I studied native prairie plants in the small, wild pockets along railroad tracks or roadsides. I recognized the same plants in big swathes of Alabama's Black Prairie when I hunted at Thanksgiving. The prairie connected my worlds—both a symbol of a wild, natural ecosystem and a familiar group of herbaceous friends. College also echoed Alabama in providing me a sense of independence as a young man. I used to spend time thinking about living in

Fig. 4.3 Black prairie with encroaching juniper trees

Greene County, even into my twenties, and I would've liked to move onto the property they called the Willis Tract and to try farming.

My friends' dad differed from the outdoorsmen I read about in homesteading books. Despite owning two thousand acres, he did not seem to be hunting for sport or for food, he was editing the wildlife populations because he had the power to do so, or because it seemed to benefit him in the moment. To me, it seemed connected to how they harvested trees and treated the land, planting the black prairie with cedar trees and cutting hardwoods from the bottomland. It didn't make sense ecologically, but didn't matter; deer would come and eat the honeysuckle growing rampant under the pines as well as they ate acorns from the big old oaks.

"How do deer fit in ecologically, Andrew?"

"They are foragers, eating grass and small trees and acorns, often at habitat edges. Deer have affected the landscape differently over the years, though, simply based on their numbers."

White-tailed deer, a species existing in North America for four million years, once fitted appropriately into their niche, kept in check by predators and indigenous hunters. These deer became rare in both Alabama and Pennsylvania—and the whole Eastern U.S.—one hundred years ago, nearly extirpated by deforestation and overhunting. With few deer to browse tree seedlings, forest-replanting efforts thrived in the mid-1900's. Then, restocking of deer also succeeded, fueled by the young forests that now lacked predators. The deer populations exploded.

Currently, there are about thirty deer per square mile in both Alabama and Pennsylvania, and, in the less-populated Alabama, that translates to about one deer for every three people. Although they are native wildlife, white-tailed deer populations are tightly linked to human activity and management. They have become pestilent for both forests and farmers, and whether or not we eat them, we are connected to them ecologically.

My ideas of what makes sense ecologically grew from studies in college and, later, in graduate school, where I learned to value biodiversity, native species, and the plant communities best suited for the soil and climate of a place. Studying forest restoration in Northern Arizona trained me to consider the pieces of an ecosystem that help it to function and support diverse plants and animals. During this phase of my life, I valued wilderness above all else, hitching my ideals to leaving the landscape alone, allowing the wild.

In April, just before we got married—and just before we moved to Arizona for my graduate studies—I made a rare, springtime trip to Alabama. With no hunting agenda, I just walked through the woods, reveling in the spring wildflowers under stately hardwood forest. When I returned to hunt that fall, it was all gone. They had clear-cut it for money, and planted spindly, fast-growing loblolly pine. I would carry that image of biodiversity replaced by monoculture with me into my forest restoration work, always asking myself what makes a forest healthy.

"Did you have a sense of loss?
"Yeah. It was this amazing thing that disappeared. It was my place without being my place. I lost the illusion of there being a true pocket of wilderness here and faced the reality that it is all managed by people. I felt sad and confused at why the forest needed to be cut down. To me, the forest had more value alive than clear-cut into logs and paper. That moment stuck with me, making me want to own land and manage it differently."

The Black Belt and those expansive properties have a history of land ownership that includes the disturbing, brutal history of conquest and manipulation through slavery. And before slavery, the Choctaw Tribe hunted, fished, and farmed this region, until ruthlessly pushed westward by the U.S. government. On my friends' property, we hunted over abandoned cotton fields. We sat in a place called the mule barn, where slaves once tended the mules. Although slavery was recent history, inequality has continued. White people, including my friends, continued to own most of the land. Most black people living in Greene County did not have access to hunting these huge tracts of land, many of which were privately-owned hunting clubs requiring memberships worth thousands for the privilege of hunting there, typically for antlers, not the meat that some families could use.

On a small knoll in one of my friends' properties sat a group of concrete block houses, the remains of a small community once served by electricity and the postal service—land once occupied by multiple families, now owned by one. Some of the houses had only a chimney and foundation. One cabin still had walls, but needed roof work, doors, and other patching. My brother moved down there for several months, beating back the rats and beginning cabin repairs. We cut and milled cedars from the place where we hunted, and—in one big project week involving our parents and our friends' family—we added electricity. Renovated, with hand-built cedar bunk beds, the cabin became a place where we felt some ownership, and we

Fig. 4.4 The cabin

stayed there at Thanksgiving. But we never lost the awareness that this was someone else's home before us, and it did not belong to us now (Fig. 4.4).

None of this land belonged to me. I only owned my experiences of it and love for it. The permission to hunt, the place to stay, guns to use, and the deer meat itself all came at the kind generosity of my friends. We rarely—if ever—discussed land use, instead spending our eight days at Thanksgiving being present in the moment. Even when our land ethics differed, I still loved them, and I owed them my supper—animals gleaned from their land.

Animals

"Abbie, Did you feel differently about the hunting when you spent Thanksgiving there during vet school?"

"I did. I felt more competent, and the insides of deer meant more to me as I helped butcher them."

Veterinary school gave me a new familiarity with animal bodies. Beginning with their cells' chemical signals and the puzzle pieces of anatomy, I learned their inner workings. In my second year, we studied organ systems—the pumping heart, the filtering kidneys, the transformative liver. By my fourth year, I was ready to put it all together and make sense of a physical examination, bloodwork, and radiographs to diagnose and treat animal disease. If this failed, I was ready to necropsy—to cut apart a patient who had died, in search of answers.

Not long after my senior year rotation through the necropsy lab, I stood again in the cleaning shed in Alabama. In necropsy, I had learned the importance of keeping your large knife sharp, and my wrists learned the choreography of sharpening it quickly. One evening in Alabama, the butcher-turned-doctor father walked into the cleaning shed and raised his eyebrows, complimenting my skills over the sound of my knife zinging back and forth against a sword-like sharpener. I joked with the hunters that the patients they had brought me appeared to have suffered gunshot wounds, and there was nothing I could do for them.

That year, no one seemed to want to bother with turning their deer into venison, and offered them all to us. We returned home with seven deer for our freezer. So despite both of us not eating other meat at that stage, we worked venison into most of our meals. One afternoon, a vet school classmate expressed his disgust at my venison-filled lunch, insisting on my heartlessness and hypocrisy while he ate his chicken sandwich. He had a point. I had become a vegetarian because, although I don't quite consider them to be human, animals are people to me, but here I sat, eating them.

In the context of my relationship with Andrew, I tried thinking differently about hunting. Eating these free-ranging deer seemed preferable to eating animals raised in crowded conditions, shipped in frightening trucks, and killed behind closed doors. If I was going to eat meat, I owed it to the animal to witness their death and to get my hands dirty preparing their body to be food.

> *"Before I met you, Andrew, I had no interest in guys who hunted. I couldn't tolerate the idea of killing animals, and the hunters who bragged about it. I hated guns. But I liked you anyway."*
>
> *"It's amazing how love is blind."*
>
> *"Or something. You seemed different about hunting. You were quiet about it, and always respected the animal and the meat. It seemed more like a natural extension of your closeness with the woods."*

My conflicted feelings about hunting in Alabama and eating venison did fit into my evolving understanding of life's built-in contradictions. I was willing to accept some cognitive dissonance, as long as we kept asking questions and aiming for a healthy love of the world. Wielding the knife to butcher our food seemed to have integrity compatible with using my hands to diagnose and treat animal disease. Some part of me will always feel sadness at eating meat—even though it tastes good—and that sadness coexists with the other aspects of our complex relationships with animals.

Game

> *"It is true that hunting there was different from other places, especially inside the fence."*
>
> *"When did they build the high fence?"*
>
> *"They fenced the Matthiessen Place, 3200 acres, in the late 1990's."*

For many landowners in the South, hunting has evolved to mean planting huge green fields to attract animals and building comfortable deer stands with carpet and closeable windows facing the green fields. Once the fence was built—ten feet tall all the way around—the deer population was contained and managed. At several feeding stations, they set up game cameras to photograph the deer, keeping track especially of the heavy-antlered bucks, which local hunters would pay a premium for the opportunity to shoot. They kept the game photos in an album, poring over it to compare racks and identify individual deer.

We were allowed to shoot does, but no bucks, and the uncle who owned the Matthiessen Place relied on our judgment to keep us from accidentally bagging a "button buck," with his young, bumps of antler hard to see at a distance. Being the key predator managing deer numbers and balancing the herd's sex ratio, the landowner needed to curb the population's potential to outgrow the food supply. One year he killed eighty does inside his high fence.

They didn't need or even much want the deer for food. Over the years, our family ate more meat from their property than they did. In a strange way, these deer were more livestock than wildlife, except the herd was managed for big antlers, not big muscles, which seemed a luxury use of the land's energy.

The relationship with deer became even more domestic when the uncle agreed to allow a Tuscaloosa-based wildlife rehabilitator to release bottle-fed fawns on his part of the fenced property. These rescued deer got purple ear tags to identify them—in theory—as deer not to shoot. Most of them assimilated into the general deer population, but a few hung around the house, where another friend from Pennsylvania was living. This friend rarely changed his crusty overalls or zipped his fly, and never shaved his unruly beard. His soft spots included Abbie and I, and the half-tame young deer. Alone with his already-dubious housekeeping, this friend started specially feeding one little buck, allowing him indoors and naming him Itty Bit. Soon Itty Bit drank from the toilet and enjoyed apple pie from a spoon. He grew antlers and testosterone, though, causing the rest of us to worry (Fig. 4.5).

Over a year later, the uncle sat in a deer stand with his son and pointed across the field.

"I don't like the way that deer looks. Why don't we shoot it?"

He was referring to the uneven antlers, which he preferred to cull from the genetic pool of his deer. His words were not a question, and his son shot the deer. When they walked out to the field to inspect the buck, they found it was Itty Bit. To this day, we allow some vagueness about whether or not the uncle knew that odd-looking deer was his friend's increasingly dangerous pal, Itty Bit.

This uncle, a law school graduate married to a judge, managed his deer closely, employing two local young men to help tag each deer at about one-and-a-half years old. Known for their sharpshooting skills and delight in shooting anything, these two brothers could dart the deer with tranquilizers at a distance. The deer wore different-colored ear tags to denote age and other characteristics, such as their origins as a rescued fawn.

The high fence enclosure and tagged deer herd seemed in conflict with my ideals of wildness and harvesting game from the landscape. Had the wild been tamed? I worried about a culture of dominance in managing the land versus one of coexistence, cooperation, and respect. I wondered how I would manage the land—and still use it to earn a living—if I owned it (Fig. 4.6).

Fig. 4.5 Itty Bit

At this phase, my wilderness ideals were commingling with respect for human efforts in land management. Deer live in complicated systems already altered from wilderness, especially in the Eastern United States, and we are part of those systems. I had begun to value a cultural landscape. My ideals began to include humans managing nature to get what they need, while working towards healthy, diverse systems of plants and animals. If I owned their 5,000 acres, I imagined sustainable agriculture and forestry. Although my friends owned the land for fifty years, they seemed consumptive and disconnected from certain ecological nuances of it. Their land management, though, allowed more wildness and function than if they had tilled the acres to plant corn or cotton, and they did not spray fertilizers or chemicals, so their venison could be considered organic and free-range.

My friends were also deeply generous. With the deer management producing a large excess of dead does, Granny established a foundation to pay for processing the deer into venison at a local butcher in Eutaw. The meat was given away. They also embraced me—and my parents, wife, kids, and father-in-law—as family, inviting us to join them for Thanksgiving dinner, stay with them, hunt, and take home venison for twenty years.

Friends

"When we got married, I felt almost like I had two sets of in-laws—your family, and your friends' family—and two equally important places to visit—where you grew up and Alabama."

"It's true—you did."

Fig. 4.6 Exceptional deer stand

One of the last years we went to Greene County, our kids—a boy and a girl—were just old enough to sit on the uncle's backhoe and toddle around the house admiring the taxidermy. Our friends also had young kids, and this generation of parents—lacking their parents' and grandparents' doctor and lawyer salaries—had a heightened interest in hunting for meat. Now that we had children, our friends especially wanted to facilitate Andrew's hunting success, to make sure we left with venison.

One afternoon on the day after Thanksgiving, the group preparing to hunt filled the house on the 3200-acre, fenced property, where Itty Bit used to drink from the toilet. Burly bucks posed, glassy-eyed, on all the walls, staring over crouched bobcats and a coyote. During the hunting hours, the children rolled balls across a pool table after the guns had been removed from it. The moms, including Abbie, watched the sunset, drank tea, and listened for shots. The dads, including Andrew, and the younger adult siblings and cousins split up into various stands, hoping to bring home the backstraps.

The hunting enterprise had often been intergenerational, from Andrew's early years with the granddad through hunting with the uncle and his friends. Andrew's dad joined him one year and shot two does and felt grateful to take home the extra meat. The very last Thanksgiving, in 2013, Andrew took Abbie's dad, who declined to carry a gun, but dove right into the butchering, which impressed him with the amount of exhausting work required to disassemble several whole deer into readily cookable portions.

In 2012, the last year we brought our kids, they were still awake and playing after dark, when the guns returned with deer on the backs of pickup trucks. They saw the deer and asked about the activity in the cleaning shed—part horrified, part entertained. We had moved beyond freezing whole shoulders in bags, and spent hours carving tough parts from tender, and cutting huge hindquarters into chunks to grind up, yielding many bags of easy-to-use ground venison. The cleaning shed was full that year, with the whole family preparing meat. The two of us took turns, swapping childcare and butchering, since at that stage the two tasks were mutually exclusive.

We found ourselves—ecologist and veterinarian—once again in the midst of dead animals that were wild yet contained on a landscape that was wild, yet managed. We worked alongside friends who had grown even more dear to us over the years, even as we had grown to be even more different people in our values and beliefs. Although there was plenty of opportunity for all of us to declare each other incomprehensible, we continued to join each other in what had become ritual. As with many rituals, our hunting and butchering of deer happened at the same time of year, same place, with the same tools, and it gave us a sense of meaning and togetherness. Our bodies remember Greene County in the ceremonial movements of gathering meat and the nourishment of the meat itself.

Looking back across twenty years of hunting in Alabama, we are surprised by its integrity, despite our awareness of and participation in its paradoxes. The experience was always a complex jumble of harvesting the year's meat amidst Andrew's love of the wild landscape, Abbie's love of animals, and both of our love of the people who invited us there. Since acquiring our own upstate New York property, we have embarked on new rituals for hunting, missing our friends each year at Thanksgiving.

Now, we are landowners, planting 600 trees last year on our dilapidated 53-acre farm and watching the overabundant white-tailed deer nibble our saplings. In addition to rehabilitating our 200-year-old barn, we grow a large vegetable garden and raise chickens for eggs and meat, swapping meat chickens with neighbors for the pastured pork and beef they raise. Some of our questions look different now, as we wonder how best to promote farm and forest health under the onslaught of wild herbivores and how to raise food and children sustainably.

We have both changed over the past decades together. We notice parallels in our journeys from adherence to ideals—wilderness, vegetarianism—towards embracing paradox, trying to manage landscapes carefully and eat meat responsibly. In some ways, the joke is on us, as we scheme about planting trees in rows for a purpose—fruit, nuts, fuel—and raising animals that we know we will kill and eat. We find ourselves, as landowners, more similar to our Alabama friends than we would have predicted. Aside from a small woodlot, our farm is mostly rich pasture, offering a wealth of grazing for both domestic and wild herbivores. We plan to fence these acres for rotating grazers, chuckling at finding ourselves not with strict preservation goals, but with farming goals, which involve intentional changes to the land, even when farming is at its most ecologically sensitive.

Now, we view the wild herbivores as both pests and dinner. Two deer were killed by people on our property this year. One spotty fawn, hiding in tall hay, was hit by the mower mid-summer. One fat doe, grazing the same hill on the last day of hunting season, fell to Andrew's rifle—the first deer he shot outside of Greene County with his own gun on our own property. While Abbie was working at the emergency veterinary clinic, Andrew butchered the deer, sending photos of our kids turning the meat grinder's long handle. We imagined our Alabama friends grinning at another generation learning the secrets of butchering, the origins of meat.

As we continue to wrestle our own ethics, there is a reassuring resilience in the land and animals and in our relationships with people. As we turn a deer into our supper, we engage with the visceral connections between humans and wildlife, land and food, and each other (Figs. 4.7 and 4.8).

Fig. 4.7 Grinding venison

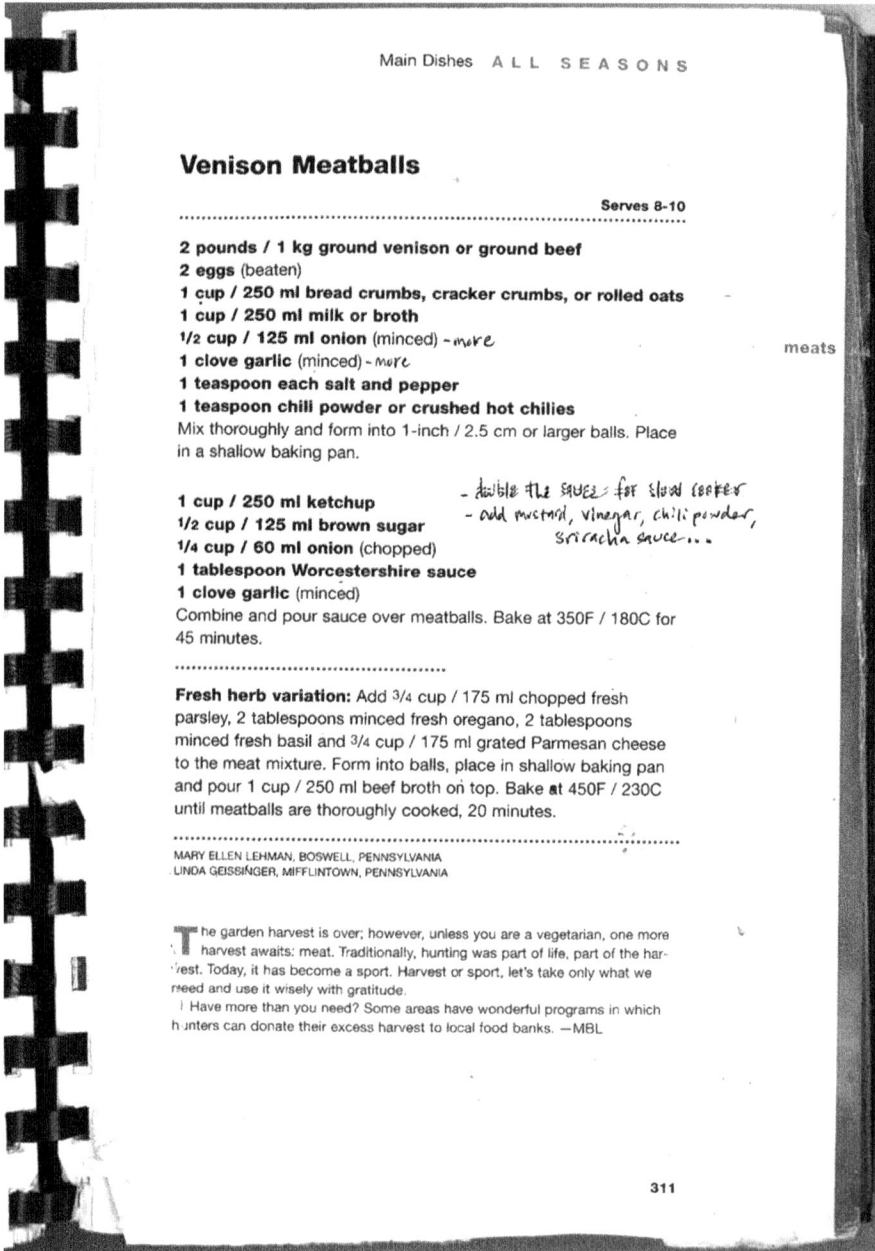

Main Dishes ALL SEASONS

Venison Meatballs

Serves 8-10

2 pounds / 1 kg ground venison or ground beef
2 eggs (beaten)
1 cup / 250 ml bread crumbs, cracker crumbs, or rolled oats
1 cup / 250 ml milk or broth
1/2 cup / 125 ml onion (minced) - *more*
1 clove garlic (minced) - *more*
1 teaspoon each salt and pepper
1 teaspoon chili powder or crushed hot chilies
Mix thoroughly and form into 1-inch / 2.5 cm or larger balls. Place
in a shallow baking pan.

meats

1 cup / 250 ml ketchup   *- double the sauce for slow cooker*
1/2 cup / 125 ml brown sugar   *- add mustard, vinegar, chili powder,*
1/4 cup / 60 ml onion (chopped)   *   sriracha sauce...*
1 tablespoon Worcestershire sauce
1 clove garlic (minced)
Combine and pour sauce over meatballs. Bake at 350F / 180C for
45 minutes.

Fresh herb variation: Add 3/4 cup / 175 ml chopped fresh
parsley, 2 tablespoons minced fresh oregano, 2 tablespoons
minced fresh basil and 3/4 cup / 175 ml grated Parmesan cheese
to the meat mixture. Form into balls, place in shallow baking pan
and pour 1 cup / 250 ml beef broth on top. Bake at 450F / 230C
until meatballs are thoroughly cooked, 20 minutes.

MARY ELLEN LEHMAN, BOSWELL, PENNSYLVANIA
LINDA GEISSINGER, MIFFLINTOWN, PENNSYLVANIA

The garden harvest is over; however, unless you are a vegetarian, one more
harvest awaits: meat. Traditionally, hunting was part of life, part of the harvest. Today, it has become a sport. Harvest or sport, let's take only what we
need and use it wisely with gratitude.
 Have more than you need? Some areas have wonderful programs in which
hunters can donate their excess harvest to local food banks. —MBL

311

Fig. 4.8 A favorite venison recipe from *Simply In Season*. (Herald Press, 2005)

Abbie Gascho Landis is a writer and veterinarian and author of *Immersion: The Science and Mystery of Freshwater Mussels* (Island Press, 2017). She has won Duke University's Center for Documentary Studies 2015 Essay Award, an Arthur DeLong Writing Award, and was a finalist for the Constance Rooke Creative Nonfiction Award. Landis holds a B.A. in English and biology from Goshen College and a doctorate in veterinary medicine from The Ohio State University. Her writing also appears in *Oxford American, Natural History, Earth Island Journal, Whole Terrain, Paste Magazine,* and *National Geographic's* blog. She relishes time with her favorite people and having lots of ideas.

Andrew Gascho Landis is an assistant professor of Environmental Science at the State University of New York at Cobleskill. He teaches Ecological Restoration, Watershed Management, Forest Ecology, and Hydrology. His research focuses on the conservation of freshwater mussels. He earned his Ph.D. from Auburn University in Aquatic Ecology, M.S. from Northern Arizona University in Forest Ecology, and B.A. from Goshen College in Biology. Outside the classroom, he enjoys raising food with his family, camping, and planting trees.

Chapter 5
Savouring the Free Lunch: Edible Activism and the Joy of Foraging

David Chang and Heesoon Bai

Oysters on the Beach

In the spring of 2016, I (David Chang) had the privilege of working on an organic farm on an off-the-grid island off the coast of British Columbia. On a mild afternoon, the sky pale with strokes of stratus clouds, I descend a steep embankment with Emily, my host at the organic farm, and alight on glistening sands along the alcove. At low tide, the waters loom far on the sightline, and the beach runs wide and clean, mirror pools flush green with kelp. I dredge through thick sand, labouring half a mile toward the water's edge, coming upon a rocky archipelago, surrounded by a sleet of shells on the sandy bed. Three of Emily's friends have already arrived, leisurely digging clams, chuckling in each other's jovial company. "The oysters don't look great." One of the women says with concern. "I guess there hasn't been much to feed on," Emily replies.

Despite the apparent diminishment in the quality of the yield, the beach looks bewilderingly abundant. Oysters strewn broadside along the inter-tidal zone – hardly a space to place my feet without crunching a shell. Emily pulls out two mason jars from her satchel and fills them half full with seawater. She shows me how to choose the best oysters – the ones that display a purple, frilled edge along the valves are ones undergoing rapid growth, and more likely to taste plump and juicy. After learning the markers, I set about foraging for prime specimens. I fastidiously sift through the hundreds of oysters in my vicinity, but only a dozen or so are big enough for the dinner table; fewer still exhibit the frilly shell that Emily prizes.

D. Chang (✉)
Simon Fraser University, Burnaby, BC, Canada
e-mail: dchangh@sfu.ca

H. Bai
Faculty of Education, Simon Fraser University, Burnaby, BC, Canada
e-mail: hbai@sfu.ca

© Springer Nature Switzerland AG 2020   55
J. B. Pontius et al. (eds.), *Place-based Learning for the Plate*, Environmental
Discourses in Science Education 6,
https://doi.org/10.1007/978-3-030-42814-3_5

After gathering a few good oysters, Emily demonstrates the art of oyster shucking. She identifies the upper and the bottom valves, jabs her knife into the side opening, shimmies the blade before prying the shell open in one deft motion. I try my hand with the next oyster. The jagged, awkward shell resists my grip and I struggle to hold the oyster against the force of my knife. I manage to penetrate the seam, but the blade is lodged tight and will not move. I steady the shell between my knees and feel the sharp, calcified crust digging into my knee. I lean into the knife in a mad push, bending, wrangling and grunting with the oyster in hand. Finally, a snap! The shell cracks open and I see my oyster, spitting sand and water, drenched in a murky fluid of gray and white. "Looks like a good one," Emily says, scraping the oyster from the shell and depositing it into her jar. She asks, "Would you eat that raw?" I scrunch my nose at the creature in the jar, bathed in slime and turbid water. Restaurant oysters served on beds of salt, arranged in a ring around the edge of the plate, garnished by lemon slices, parsley and thyme seem an entirely different delicacy. How sanitized a display compared to these uncanny creatures on the beach, unwieldy to touch, redolent of fish and sea, unyielding as volcanic rock! Pressed against the glassy jar, their smoky, corpulent flesh is the furthest remove from the longings of taste and appetite.

We fill two jars full of oysters and three bags of clams before returning to land. The tide rising again, we pick up our pace as water rounds rock and boulder. "I'll fry these with some butter and garlic – they'll be very tasty," Emily says. We return with an impressive harvest for forty-five minutes of work. I scan the surroundings again in admiration of the exquisite landscape: the dark pines on the bank, the distant waters, the outline of island slates, the assertive boulders, and tumbling burl of rock that carves the alcove. Such breathtaking beauty and resplendent abundance! Emily and her husband, Joseph, have shucked oysters for the better part of thirty years. They follow the tide table, which sometimes bring them to the beach at three in the morning, under the pale rays of a luminous moon, the entire landscape lit in a fantastical dream of blue and silver, and the cosmos wrapped in deep, enduring silence. Nestled deep in this dark and shimmering universe, they roam the sands and pluck nature's pearls.

But isn't this a deal too sweet? Mustn't there be someone looming ashore, scale in hand, and awaiting payment for these prized delicacies? Perhaps we should at least *tell* someone that we've collected our dinner on the beach, lest we be accused of thievery? Yet, without fanfare, we load the harvest into Emily's truck, hug our friends, and return home with the days' bounty.

Eating in the Age of Commodity

Since my (David Chang) oyster shucking experience, I continue to ruminate on the significance of this foraging event in relation to my sedentary, urban life. I live amidst slabs of concrete and towers of glass, where plastered walls and glossy planes rise from steel and asphalt. My foraging places are supermarkets, where the "produce" is arranged in lines, stacked in strict symmetry. My breakfast comes in a cardboard box – if I make enough space in my pantry, I can line the cereal, the pancake mix, corn starch, sugar and the powdered protein mix on one shelf – a variable collection of edibles. I survey the pantry as I would a shelf of books, fingers tracing spine. Each food item can sit on that shelf for the better part of three years without spoilage. All this plentitude is stored without the slightest odor that might incur on my living space, and the slightest demand on my attention and care. When did food become so staid, entirely devoid of imaginative possibility, and completely empty of meaning and excitement?

Immersed in the pervasive abundance of edibles produced by the commercial economy, we have traded communion for convenience, mutuality for materiality. Whereas my foraging episode on the beach encompass a suite of contacts registering on the somatic, emotional, psychic, temporal, aesthetic and gastronomical dimensions of experience, purchasing groceries in the city is an entirely different affair, one ruled by strict utility and transactional efficiency. Grocery shopping is often a chore and rarely a sublime experience. This stark contrast raises questions about the most basic of human acts: *How do we procure our food and how do these procedures shape us in turn?* For most of us raised in cities subsidized by an agricultural complex, the procurement of food is merely another monetary exchange; the ecological webs that undergird this interchange are easily obscured by economic transaction. The departure is far enough that we see *shopping for groceries* and *foraging for food* as two disparate orders, each belonging to distinct epochs in evolutionary history. Foraging is relegated to a bygone era – if not shrouded in quaint nostalgia, then scorned for its unabashed primitivism. Reliance on the products of the capitalist-industrial food complex remains the default mode of consumption; anything other is perceived as retrograde to the inevitable march of progress.

On the other hand, the image of the forager in the collective imagination may signal a perennial longing for ecological connection. The earth's primordial fecundity reveal themselves in the bounty of the forest. Consider the childhood fantasy scene of the wondrous candy-land, where rivers flow with scrumptious chocolate milk, and where lollipops bloom like flowers on rolling hills, under clouds of cotton candy. This vision of a paradise for the palate manifests a nascent fascination with abundance and an intuitive grasp of an edible world where sustenance always lies within arm's reach. In reference to the fecundity of the earth, John Milton wrote of Adam walking "Into the blissful field, through Groves of Myrrh/And flow'ring Odours, Cassia, Nard, and Balm;/A Wilderness of sweets…" (Milton 2012, p. 2031). This dream of eating the world is not a display of childish whim, less an atavistic regression to primitive fantasy, but rather a manifestation of our ecological affinity to land and its splendid bounty. Like the candy-land fantasy, human ancestors who foraged for food experienced their sustenance as gifts freely bestowed, and by extension, knew their lives to be dependent on the larger matrix of life.

At the same time, we should not indulge a sanguine notion of foraging as a placid idyll from an innocent past. Foragers, from the smallest rodents to prodigious herds of buffalo, are vulnerable to predation when they roam the landscape. Foragers must negotiate several dilemmas – the energy they secure through expedition must at least equal or exceed the energy they expend in foraging, or they will have squandered both time and strength. They must guard against predators – vigilance and fleet-footedness both exact a cost to energy reserves and constrain their ability to glean sustenance from the landscape (Stephens et al. 2008). In response to the risks associated with foraging, animals have evolved many strategies – some forage in herds, thus distribute the task of vigilance amongst a larger number of peers (Stephens et al. 2008). Others undergo a physiological transformation such as camouflage (for example, the spotted flounder) and quills (North American porcupine) to evade and repel predators. These wide-ranging adaptions instantiate the grave threats posed by foraging and the need for a repertoire of strategies for avoiding peril. Further, dependence on the biome means vulnerability during times of scarcity. Prior to agriculture, human foragers were also buffeted by inclement weather and unpredictable terrain. Such are the risks of dependence on a dynamic biosphere; gratuitous abundance is never far from peril, and life skates close to the clutches of death.

Since the end of the Pleistocene, many human civilizations have relied on agriculture as their main food source, although the foraging tradition remains strong in many indigenous communities. For those who live in large metropolitan centers, however, foraging is no longer (and cannot realistically) serve the primary means of procuring food, because cultural knowledge of edible flora has declined, and bands of wild spaces have diminished. If foraging offers no realistic solution to the current demand for food, of what relevance might it pose to an increasingly urban populace acculturated to mass industrial agriculture and the economy of monetary exchange? What educational and cultural value might we glean from the practice of foraging, and how might it transform sensibilities and initiate participation in knowledge traditions conducive to a comprehensive ecological ethic?

In response to these questions, we present a conversation between the authors of this chapter: two city dwellers better apt to navigate the polished isles of the nearest Whole Foods than the tortuous trails of a boreal forest. We reflect on our foraging experiences and explore its significance to our urban lives. Having been born in Asia (David in Taiwan and Heesoon in Korea), we bring our views on food, which we inherited from our past, to our hybridized cultural present in a modern, Western society (Canada). We shall explore the educative potential of foraging practices, with attention to the sensual component of deep engagement with wildness, the experience of which returns our somatic, mental and spiritual connection with a numinous earth. We highlight the cultural and relational knowledge traditions that are indispensable to foraging practices, and the vital role of reliable guides who initiate us into an edible world. At this intersection between ecological connection and cultural knowledge, we forward some reflections on how foraging practices might constitute forms of social and ecological activism within the capitalist-industrial food complex, a contrapuntal narrative to consumer culture. Thus we present foraging practices as a form of edible activism, and consider how foraging may facilitate an ecological consciousness via practices that imply participation in a *sacred economy*, where gratitude and delight accompany the cycles of life.

Sensuality and Somatic Attunement

David Chang (DC): Heesoon, we both live in the city, near downtown, and both in apartment buildings. We are in the bowels of a human-made world, so foraging seems like an entirely alien concept. So I wonder about your experiences with this seemingly strange and foreign activity. What experiences have you had with foraging?

Heesoon Bai (HB): I grew up in Seoul, the biggest city of Korea, in the 50's and 60's. In those days, we didn't have supermarkets. Not even "groceries," as we know them today. We had sprawling open markets that stretched many blocks in which vendors laid out, mostly on the ground, their items to sell: live chickens in cages, pans and barrels of all manners of seafood, farm-grown vegetables, and foraged mountain vegetables that were in season. As a child, sometimes I would accompany my mother on her daily trips to these markets (we had no refrigerators those days), keeping myself closely attached to my mother, for a sense of safety and protection. I found the sight, sound, and smell of the market often overwhelming, and certain parts, especially places where land and sea animals, dead or alive, were displayed and sold, somewhat frightening. My mother was very confident and skilled when it came to handling merchants: in the main, these folks were not the politest lot. Some of them were deceptive, but my mother knew how not to fall for their guile. And she was fearless and confident – to my young eyes, terribly brave – when it came to handling edible items, both vegetative and animal. Vegetables were soil-covered, and dead animals looked pretty frightening to me. There were all sorts of strong aroma, scents, and fragrances from fruits, leafy vegetables, earthy root vegetables, and seaweeds, as well as strange smells from live and dead animals and decaying matter. And my greatest admiration went to my mother's skill at turning these wild-looking materials into delectable dishes. Preparations involved a lot of physical exertion outside the kitchen, in the courtyard at the water pump. Removal of soil from vegetables, especially mountain vegetables, was a routine step. As I grew older, I tried to help out with these preparations, including a major winter fish freeze-drying activity that involved gutting hundreds of Pollock in icy water and hanging them on a line for some weeks.

With this childhood background, and having arrived in Canada in 1972 when I was 18, I found the experience of shopping in a supermarket a most bewildering and disembodying experience. No dirt, no smell, no sound (I don't think they even had music playing in the store those days), and just about everything was neatly cut-up and shrink wrapped. And no vendors to talk to and haggle with over the price or quality of goods. I can only say that it was a surreal experience.

   Now, what is intriguing to note is that I started foraging after I came to Canada, not while I was growing up in Korea. Coming from a crowded urban jungle (even though I lived close to the mountains on the outskirts of Seoul), what I noticed in my new homeland, Canada, was the abundance of green space: lots of land all around such as large backyards, back lanes, empty lots, parks, woodlands, and so on, that were all showing the vigor of a Green Kingdom. All manners of plants, shrubs, and trees proliferated everywhere. I began to forage once my mother immigrated to Canada (during my twenties) and started to spend many months of the year with me. She started to show me the wild edible plants she could identify from her childhood memory. She grew up in rural southern part of Korea foraging wild mountain vegetables and catching fish! Not a usual activity for a Korean girl in her time, but thanks to her father, she did everything to her heart's content what boys would do. Most common weeds (for us they were "mountain" or "wild" vegetables) my mom and I harvested in Canada were "pigweeds," "shepherd purse," "bracken," "dandelions," and "broadleaf plantain." I learned to prepare them the way my mother did, blanched and seasoned with crushed sesame seeds, sesame oil, soy sauce, finely chopped green onion, finely minced garlic, and hot pepper flakes. At some point, I read up on these wild edibles and found out that their vitamin and mineral contents were substantially higher than cultivated vegetables with a similar taste. I didn't have much money those days (living below the official poverty level), and so I felt very smart about my frequent foraging practice. More importantly, however, I really appreciated the bonding experience I was having with my mom. We were both too busy to spend time together while we were living in Korea: she with struggles to take care of our large, complex, and challenging family, and I with my own survival struggle of doing well in school, which culminated in surmounting the final hurdle – the University Entrance Exam. It was a pleasure indeed to leisurely roam the neighbourhood with my mom, gathering wild vegetables. It was also very meaningful to me that I was connecting to my mother's roots: to her rural and wilderness-rich childhood and to the forgotten and rejected "pre-modern" ways of life that were closer to soil and all that was wild and natural. Foraging was second nature to my mother.

DC: That's a precious educational experience you've had with your mother. I can gather that you are much more knowledgeable a forager than I am. I am surprised that your mother's knowledge of edible flora applied to the west coast of Canada, a different bioregion. Do Canada and Korea share some of the same wild edible plants?

HB: I haven't done research to see how on earth these two regions share many such "weeds," but they do. My guess is that the same "wild vegetables" (so called "weeds") my mom identified in Canada belong to the same genus but different species. This is the case with, for example, trees. We have pines in Korea and pines in Canada: they may belong to the same general type, genus, but to different species.

DC: I see. Your story reminds me of my own childhood experience. As a kid in Taiwan, I hadn't had much exposure to green, wild spaces. I remember my parents taking me for a long drive to the countryside on a Sunday afternoon. We came to a large stream where a group of family friends had gathered. My brother and I were given dip nets and told to wade into the water to catch fresh water smelt. I was astonished to see schools of fish moving from one pool to another in that coursing stream. Pants rolled above our knees, we chased the fish and laughed in exhilaration. My brother and I developed strategies for netting the fish – I would herd a school into a shallow pool while my brother waited on the other end where the frantic fish would exit. I dug my net deep in the water and drove the dashing streaks of green and silver toward my brother's steady guard. But the fish were quick and the nets heavy and awkward in water. Our strategy only caught us a few small fish, but we were entirely undaunted. The sheer joy of the task was enough. Combined with the efforts of others, we managed to catch enough fish for a feast. My parents called me in for dinner, but I lingered in the stream, fully immersed in the game. Dinner was scrumptious – fried smelt dressed with cilantro and chopped onions. I had worked up an appetite after an afternoon in the water. Prior to that experience, I'd never understood playing and eating as the same activity. In that afternoon of catching smelt, I relished the stream, and the delectable taste of the fish in one indelible experience. I didn't know it was possible to have this much fun catching my own food. It was an entirely delightful discovery.

HB: Exactly! That was my experience of foraging, too: pleasure all around. The most prominent foraging practice that my daughters and I cultivated was picking berries. My younger one was an older toddler and my older one was seven or eight, and we would be picking wild black berries (and other berries like salmon berries) all throughout summer. We would do this every summer, and we usually ended up with around over 20 kilograms of sweetest and juiciest blackberries frozen in our freezer for winter. My seven-year old was incredibly focused and good at picking big and sweet black berries without getting hurt by the thorns. She would fill bucket after bucket. My toddler was on the ground, continuously sampling picked berries, grinning wide with red-purple stained mouth, and proudly displaying her two hands, each fingertip sporting a berry cap.

DC: I see a wonderful tableau in my mind's eye just listening to your story. What a delightful picture! Now, what significance, educational and otherwise, do you find in those memories of picking berries with your daughters?

HB: *Biophilia* – love of Nature, comes to my mind right away (Bai et al. 2010; Wilson 1984). Foraging trips took us outdoors a lot, which grew my children's love of forests and all places green and wild. While they are urban dwellers like myself, their love of and respect for Nature is very strong and prominent. While they don't formally work as environmentalists, they have the environmentalist consciousness. On a more practical note, having learned to forage, they have, I would say, a different sense of relationship with food and sensibility around what they eat and how they prepare food. They are in their early thirties and late twenties, and over the years, I have noticed that they eschew processed food and choose edibles that come in as wholesome a state as possible. And another thing I noticed is just how much familiarity and 'handiness' they show with respect to food and how they prepare them for eating, which is like me. I am never at a loss when it comes to getting hold of edible materials, whatever these may be, and preparing them for eating. I have the confidence, although moderated at this point by reduced physical stamina that comes with aging and two decades of largely sedentary life style that goes with being an academician, that, should I have to survive on foraging, I could do so, at least for a while. Were I in my twenties and thirties, my confidence level would be higher. As well, it would be very helpful to have more hands-on lessons, as my mother showed me, with more varieties of wild vegetables and seaweeds. Thinking like this at least gives me a sense of survival security that we don't have to solely depend on what supermarkets and grocers bring to us for survival. In fact, when I think of how completely dependent we city dwellers are for our sustenance on supermarkets and shops, and the long and delicate chain of causality that supports food industry, I feel insecure and even slightly panicky. It seems to me that the further removed we are from the source of our sustenance, the more vaguely insecure we would feel. At bottom, it's helplessness that triggers survival fear. That one can put one's senses and limbs to work to find and prepare food is empowering and promotes a sense of security.

DC: I know what you mean about helplessness leading to a sense of insecurity and fear. And I respect and admire your sense of confidence about being able to survive on foraging, but I still can't imagine how you, as an urbanite at this point, sourcing your foods from places like Whole Foods, could survive as a forager.

HB: Your skepticism is warranted. I think I need to comment not directly on foraging per se but on poverty. I think my sense of confidence is derived more from my experience of witnessing as well as experiencing poverty than just from my foraging experience itself. Although I have not experienced living in abject poverty, I lived, in many phases of my life, in proximity to poverty as a child in Korea. As well, I discovered at one point in my immigrant life that, to my surprise, my young family and I qualified as "members" of the governmentally declared category of poverty in Canada. This was when I realized that I was a person who was very resourceful when in difficult situations. So, while I enjoy and appreciate my current relative wealth as a university professor and my expanded purchasing power, my own repertoire of life experiences includes a wide range of experience from poverty to upper levels of income, from survival to luxuriant flourishing. Like a performer, as I described above, I feel that I can perform survival, if necessary, and I think I can put my survival skills to work, should I face survival challenges. Again, I have to attribute much of my own self-belief and faith to my mother who exemplified a life of extraordinary survival: bearing and raising five children not only under conditions of poverty but also in mortal danger during the Korean War.

DC: I sense in your description a paradox of precariousness and joy. Your mother foraged because she had to be resourceful amidst the challenges of the times. You also learned to forage because you were living close to the official poverty line. Foraging was a joyful activity practiced in the midst of those precarious moments.

HB: Yes, necessity can call forth resilience and resourcefulness. When we adapt to the situation, the situation changes and we can slowly uncover the joy that lies within.

DC: I think the joyful part of foraging stems from its sensuous dimension. Thinking again of my time catching smelt, wading through the cool water, leaning against rocks, feeling the pebbles under my feet, discerning texture and scent – this was an immersion in a world rich with sensation. The involvement of the body, lunging into a world of movement, nuance, and colour, was an enthralling adventure for a boy who had spent most of his time in classrooms. It was a feast of the senses before I even tasted the first morsel of fish. One is gratified in many ways – we don't know the hunger of the senses until we stumble upon a sumptuous boon that awakens our bodily longing.

HB: Ah, yes, bodily longing! I know a few people around me who avowedly (and maybe proudly) say that they only eat to live. I have joked that I probably could put pieces of cardboard in front them, and they would eat, as long as I drench them with enough spicy sauces. My joke aside, this lack of 'intimacy' concerns me in that, to me, it's intimate knowledge that brings about relationship of love and respect. The interaction we have with food is an intimate, and possibly sacred, act. Indeed, what can be more intimate than putting parts of the world right inside you? The ingested food then goes through a long process of being intimately handled, which in its process nourishes you. When I contemplate this process of interaction – why not call it edible love-making? – with the plant and the animal kingdom, I am overcome by a deep sense of awe, wonder, love, and gratitude. I love these lines from Dogen Zenji (1200–1253), the founder of Soto Zen in Japan:

> Handle even a single leaf of a green in such a way that it manifests the body of the Buddha. This in turn allows the Buddha to manifest through the leaf. This is a power that you cannot grasp with your rational mind. It operates freely, according to the situation, in a most natural way. At the same time, this power functions in our lives to clarify and settle activities and is beneficial to all living things. (Dogen and Uchiyama 1983, pp. 7–8)

My practice of foraging and eating is an encounter in sacred love. Of course, there is the question of actual practice and sustaining it. Practice requires commitment and discipline of life-long cultivation. I can *fall in* love for a short while but *to be in* love is a serious life-practice, requiring daily self-cultivation. I am working on it every day, noting my struggles, learning to love the struggle, learning to love the self that struggles, watching the struggle becoming a play, and so on.

DC: Your idea of *edible love-making* speaks to a kind of intercourse – the passing of life substance from one body to another, a sacred exchange, like you mentioned. I like Gary Snyder's poem "Song of the Taste," which illustrates the confluence of sex and eating as parts of the same life-affirming act: "Eating the living germs of grasses/Eating the ova of large birds/the fleshy sweetness packed around/the sperm of swaying trees… Eating each other's seed/eating/ah, each other./Kissing the lover in the mouth of bread:/lip to lip" (Snyder 2007, p. 3).

I read in Snyder's poem, the union of *consumption* and *consummation,* eating and sex as a co-extensive act. In *Practice of the Wild,* Snyder writes: "To acknowledge that each of us at the table will eventually be part of the meal is not just being 'realistic.' It is allowing the sacred to enter and accepting the sacramental aspect of our shaky temporary personal being" (Snyder 2010, p. 19). This exchange of bodies through the act of eating is a fundamental form of mutuality and reciprocity that furnishes the basis of ecology. Thus, *"kissing the lover in the mouth of bread"* reinterprets Dogen's instructions on handling the leaf of green as the body of the Buddha. There is something imminently numinous and tender about this consummation: we don't just ingest a meal, we assimilate the universe with each morsel, and we in turn season ourselves as nourishment for others.

HB: So sublime and yet so humble!

Culture and Relationship

DC: Now I'd like us to discuss the cultural and relational dimensions of foraging. It seems that few people become successful foragers entirely on their own. There's a vast body of knowledge that one must tap into before one goes out to gather food; usually there's an experienced expert who imparts knowledge of edible flora to a novice forager. You had your mother who imparted her knowledge of edible plants. I wonder if many people have access to this traditional expertise, and I worry about the impoverishment of our collective knowledge when this affinity with the foodscape falls into further decline.

HB: I share your worry. In fact, regrettable to say, I am a product of that impoverishment of our collective and, I would add, Indigenous, knowledge you speak of. What I learned from my mother is only the tip of iceberg in her vast embodied knowledge bank. I have the deepest regrets that I didn't manage to do more exploration with and learning from her. Basically what took place in my family, and in my Korean culture at large, is modernization, which is synonymous with Westernization, and devaluation and marginalization of Indigenous embodied knowledge, such as my mother had. She too didn't know, in any full and explicit measure, its irreplaceable value, as she aspired to have her children join the wave of modernization. She succeeded supremely in her effort in that all her children ended up in North America, earning doctorates and becoming professors, surgeon, engineer, etc. Today, we in Canada, as elsewhere, are aware of what took place under colonization and the spread of modern Western epistemology, with consequent delegitimation of Indigenous ontologies and epistemologies, and how this has impacted negatively on both the social and the biotic spheres of wellbeing.

It is not just with my mother that I lost the opportunity in finding a guide to a body of knowledge that's alternative to the modern West. My father was an accomplished Traditional Oriental Medicine (what we call these days, Traditional Chinese Medicine) doctor, but he didn't have the chance to transmit his knowledge and skill to his children. Encouraged and motivated by my parents, my siblings and I all left Korea as young people in pursuit of "better," that is, more "successful," knowledge systems and life opportunities. I am now in a position to think about all this in terms of my relationship with my two daughters and other younger folks around me. At least I managed to show my daughters to identify pigweeds (Amaranthus retroflexus) and prepare them for a delicious dish, but I want to do much more! As an educator, I want to help people to see that success in life does not equate with dollar signs, that wealth, health, and wellbeing could (and should) have a different feel and look than how it appears in the heart and face of late capitalism of today where everything, even living beings, is commodified and instrumentalized. But I also realize that what's behind the face of capitalism is survival fear and anxiety, and that rebuilding an embodied sense of security through recovering our existential bonding with nature and world/life.

DC: And there's so much to do if we want to address the follies of capitalist society. In some ways it's not about inventing something new, but recovering something old, developing a neglected potential. It's astonishing to see what dramatic cultural changes can occur within two generations – the accumulated knowledge of many centuries, indeed millenniums, can be eradicated within a few decades. We need people through whom we can access traditional knowledge. I consider myself a child of the modern industrial era, with no strong attachment to local, place-based traditions. Up to a few years ago, I didn't have anyone in my social circle who I count as a naturalist. However, I was very interested in becoming a naturalist and was captivated by the idea of being able to identify nourishing foods that grow in the wild. I obtained a copy of National Geographic's field guide to North American shrubs. I took the book with me on walks through local nature reserves and tried to identify the berries along the trail. Having come upon a purple berry, I leafed through my field guide and found a close match – I believed I was looking at an elderberry, but the illustrations on the field guide for the pokeberry seemed awfully similar. I couldn't be sure, since the plant seemed to be young and did not have the full branch structure that was illustrated in my field guide. Having read *Into the Wild*, the cautionary tale of Christopher McCandless, who died in the Alaskan wilderness from eating a toxic plant that he misidentified (Krakauer 1997), I shuttered at the thought of making a fatal mistake in consuming a poisonous berry. Without the help of an experienced forager, I could not be certain of the plant. Books do not respond to questions; they only repeat the same information in every circumstance. Thus, I was left to struggle in ambiguity. Erring on the side of caution, I left the shrub and relinquished my autodidactic naturalist education. When my own health was on the line, I could not place complete trust in book knowledge.

Michael Polan writes of a similar conundrum in *The Omnivore's Dilemma*, when he comes upon a pod of fungi that he identifies as chanterelle mushrooms. However, without verification from a learned forager, he could not be sure. His guidebook did not instill absolute confidence, so he threw out the mushrooms. It was only later, with the help of a mycophile, that Polan was able to confirm his find – they were indeed chanterelle mushrooms. The apprehension that accompanies a harvest of mushrooms unfurls from the ominivore's dilemma: Although humans are able to eat a wide variety of plants and animals, not all plants and animals are safe to eat, nor are they all ethically unproblematic (Pollan 2006). For example, the turn to agriculture brings a host of ethical challenges: clearing forests for crops, diverting water for irrigation, raising and slaughtering domestic animals. On the other hand, absent adequate knowledge, the forager is liable to get sick or die from the food he gathers. Pollan underscores the inherent paradox that ensnares the omnivore: the ability to eat a variety of foods is not always a boon; rather, it introduces many layers of complexity into human life.

The role of an experienced guide, therefore, seems vital to a forager's education. We need the discerning eyes of experience to root our knowledge of local flora. Printed words on a page do not suffice. We need to check our perception with the expert's trained eye, the assurance of watching the expert chew on a leaf or fruit in order to confirm the veracity of our newly acquired knowledge. The forager's education is thus inextricably relational.

HB: Yes, that's right! "Printed words on a page do not suffice." Abstract knowledge, devoid of human breaths and flesh, can't guide food practices in any substantial way. I think this discussion on foraging has a far-reaching educational significance beyond foraging for actual pigweeds, mushrooms, or oysters. They are concrete instances of – I would not hesitate to say – Indigenizing curriculum and pedagogy (Manulani 2008).

DC: Indeed, foraging is a practice, and should not be reduced to a knowledge domain. I also wonder if you can elaborate on what you mean by "indigenizing curriculum"?

HB: Right. I have to admit that I feel slightly nervous to use the term, "Indigenizing curriculum," for potential misunderstanding, as the current context for this well-recognized term in education is Indigenous people and their epistemology, and it's not my intention to insinuate myself and my discourse in their work for Indigenizing education. Rather, my mentioning the term is to affirm the epistemic importance and relevance of Indigenous curriculum, and also to align myself with the larger thrust of educational awakening (Nakagawa 2000) and movement that rightly see the damaging effect of the modern Western colonial worldviews and values on the planet and its ecosystems (of which humanity is part). Under these worldviews and values, local land/sea-based and place-honouring sacramental ways of knowing and doing became marginalized and devalued. Hence, going forward, recovery of Indigenous curriculum is about learning that takes place in the particular and concrete relational matrix of teachers and students who are interested and concerned with local phenomena, issues, problems, and matters of significance and meaning (Greenwood 2013). "Local" here does not mean that it is separate from and untouched by "global." Rather, teaching and learning should insist on working with the concrete particulars of people and place, their real joys and pains, passions and concerns, in the living matrix of their relationships (Gruenewald 2003).

DC: What you say makes total sense in light of our respective experience of learning to forage. Going further, given that you learned to forage from your mother, and then proceeded to pass that knowledge to your daughters, and I learned to shuck oysters from Emily, what we are saying here reminds me of feminist epistemology in which the abstract is aligned with the masculine, and the concrete particular, with the feminine. It seems to me, then, perhaps the very act of learning foraging, with the relational bonds that it foments, is an entry to feminine epistemology and ontology (Witbeck 1983), something that I think the world really needs.

HB: Oh, I love where you are going with it! But let us be clear that the feminine and the masculine here are not to be identified with sex and gender but modality or archetype, like the Chinese concepts of *yin* and *yang*. I do stand by the understanding that the world suffers from yin-yang imbalance, dangerously skewing towards *yang* (Thompson 1996), and hence that foraging as a feminist epistemic practice would be an important healing practice for our imbalanced world.

Education and Edible Activism

DC: I'd like to complete our conversation with some considerations about how foraging might relate to a discussion of ecological ethics, especially if we recognize that current state of industrial agriculture, and the mass-production of food-like products by multinational corporations. Food has been overtaken by capitalism, and it doesn't seem realistic within the present milieu to assume that we are all going to become foragers. If that's the case, then what role might foraging play, even if a limited one, within a broader project of environmental education? You've already mentioned that foraging is a return to an Indigenous knowledge. Your experience with foraging in the middle of a Canadian city also shows that we don't need to travel far into remote wilderness to harvest the bounty of the land. So how might this kind of education counter the egregious effects of a capitalistic food culture?

HB: Wonderful question, David! Thank you. I love to muse that everything we do has both literal and symbolic significance and implication. So, I think, foraging has a deeply symbolic significance (besides the literal significance of, for example, foraging oysters and feasting on them) in that it could shift our consciousness when we engage in this activity. This shift may occur along the line of feminist, Indigenous, ecological (and so on) epistemology and ontology. If this shift could be pursued with passion and devotion, and *en mass*, it may create a tectonic shift in the matrix of the current civilization. At least that's the hopeful vision I entertain. If you ask me how likely this will happen, I have no answer! However, if a vision is worthy, then we should (therefore, I will) pursue it regardless of its outcome. What do you think?

DC: I'm with you in your envisioning. Now, to pursue this shift in the concrete particular example of food production and consumption, I think the industrial food system obscures the ecological context of our food and insinuates a pervasive habitus (Bourdieu, 1990) in which the physical and somatic dimensions of food preparation is rendered irrelevant to eating – we bypass the labour, the immersion in dirt and grime, and the elaborate ritual of gathering and preparation a meal. When food is easy, we negate the wholeness of the ecological context encompassed by the act of eating. For instance, when I pull a cluster of mushrooms from the trunk of an alder tree, I see the interdependence of the fungi and the forest landscape. The oyster shucking experience put me square in the oyster's habitat, deep in the sand, the rocks, the cold ocean water. All of these experiences circumscribe the contexts from which food springs. The mushroom has its own home – a space of relations with other organisms that support each other's flourishing, a space in which I am a guest seeking sustenance. These ecological spaces demand a quality of awareness and comportment. When foraging in the forest, I prime my senses and sharpen my attention; I soften my body and negotiate the woodland slopes with care. Attention is the price of admission. I must undergo a certain adjustment to my state of being in order to traverse into this ecological space – the land demands a way of being, enlisting me in a specific form of participation. I must submit to the land in exchange for its bounty. Prolonged engagement in this kind of activity sees me becoming a creature of the forest, the ocean, the marshes.

The supermarket, on the other hand, abrogates this participatory exchange. We walk the aisles and browse the shelves for goods that are always present to our desire and whim. They do not call upon any somatic or mental posture, nor do they obtain any sense of context or relationship with a larger whole. Phenomenologically speaking, the supermarket enacts a peculiar order: goods do not have their own context – they are units that compose an artificial environs in which the human finds her every need and whim gratified without the least demand on her person. The supermarket seems to say that everything serves human convenience, that nourishment and delectations are merely a matter of course. I think this kind of distortion of ecological relationships can be detrimental to our identity as an earthly species; it disposes us to an expectation of ease, makes us adverse to the ecological mutuality that demands an exchange of physical labour for sustenance. To this end, I think foraging experiences can provide an insightful contrast to our engrained habitus, and our acculturation within the artificial purlieus we have constructed, which reinforce the centrality of the human in a more-than-human world.

Having had a few experiences with foraging, I tend to notice the plants around my neighborhood more. Recently, I noticed mushrooms sprouting from the lawns around my neighborhood. I took some photos and cross-referenced them with information on the internet. I never went to pick the mushrooms, but my eyes have started to see the streets as an eco-system, with all kinds of life peaking through the soil. I also started to look for kelp around the False Creek seawall. I've always enjoyed seaweed, but I never occurred to me that the kelp strewn around the seawall were as edible as ones served in restaurants. I haven't gathered any seaweed yet. . . something in me worries about the pollution in the water. So I can't claim to have made huge strides in becoming a forager, but it's this new attention that has opened new possibilities. I *can* imagine myself wading into the water on a remote island, gathering a handful of seaweed and laying them out on the deck to dry. This is a small step away from reliance on the supermarket, and toward seeing nature's providence, even in the city.

HB: Well-said! Foraging would be a medicine specific for this ailing civilization that has lost, literally, its senses, and mired itself in thick and deep delusion. Foraging can be seen as eating our way into a hopeful revolution. A form of activism, yes? How about "edible activism"?

DC: Sure! However, I've always thought of "activism" as a heavy word. Activists are blasted by water cannons, choked by smoke bombs, arrested *en mass*, sometimes assassinated for the challenge they pose to the establishment. However, *edible activism* can be a playful, celebratory way of carving out space for other possible ways of living in the midst of industrial-agricultural domination. Edible activism might not be thought of as an outright challenge to the status quo, a daring defiance of the prevailing food system, as if we are poised to march the streets and brace ourselves against riot police. Instead, it is a wistful, delightful experiment in reacquainting ourselves with the primordial bounty of the earth. We can start by plucking a few sprigs of spruce to make spruce tea, or wading to the island in the middle of shallow lake to taste some blackberries. It's about becoming acquainted with elders, the keepers of ancient knowledge, spending time under the sky, passing an afternoon in quiet conversation while sampling the delights of the field. In doing this, we learn perhaps for the first time that there is indeed such a thing as a free lunch, that such a meal may be more satisfying.

Foraging, as a form of edible activism, can serve as a contrapuntal narrative to the pervasive habitus formed under the auspices of the industrial agriculture complex. By "contrapuntal," I mean the counterpoint form of musical composition, where two independent melodies are played simultaneously. The melodic lines may spread apart or interweave, at times generating tension and at others producing harmony. A musical counterpoint is not a direct binary opposition, but another sonic dimension that introduces further emotional and aesthetic texture to a composition. I think of edible activism in the same way – I do not expect foraging at this point in human history to replace mass agriculture; rather, it is a way to tease, stretch and colour our experience of the earth as ecological beings. We are conditioning our views and ingrained habits toward a more expansive participation with the earth's inherent fecundity.

Of course, it's easy for me to think of foraging as a wistful counterpoint to industrial agriculture. My own life is not dependent on foraging, and I can count on well-stocked markets for my next meal. I can muse about foraging without ever worrying about hunger or starvation, so I realize there is a risk of recommending something that smacks of romanticism, making virtue from a field in which I am only a dilettante. At the same time, it doesn't take much to breach this kind of urban romanticism. The other day, a friend of mine shared some fir tea with me, using sprigs of Douglas Fir he found on his walk. We drank the tea and commented on how the aroma smelled faintly of ripe strawberries. This unexpected association was enough for me to appreciate fir trees anew. I now pause to smell branches to catch any whiff of novel scents. It's a small step toward curiosity, toward a more eco-centric way of life. So perhaps the revolution may not topple the system tomorrow, but the practice of foraging might be a kind of practice of the self, a re-configuration of our dependencies and inculcated judgements about food culture.

HB: Yes! "Free" in your "free lunch" has a whole lot of different meaning for me now. 'Free' as a verb, meaning 'liberating.' Foraging as a liberating pedagogy, free-ing ourselves from the conditioning of the industrial-consumer capitalism that has enslaved this civilization. Now, practically speaking, I am not suggesting that we get into foraging as an exclusive daily food procuring practice. Rather, I am suggesting that we experiment with and practice foraging as a way into edible activism, and invite a host of other practices related to food finding, preparing, and creating. This suggestion I'm making is first and foremost for myself. Over the two decades that I have been working in the academy, I gradually drifted off from doing a lot of the food preparations that I used to do before re-entering the academy. I am now trying to return to my food practice. I am happy to report that I'm doing a lot of different kinds of fermenting (for example, kimchi, sauerkraut, kefir, kombucha). Fermenting is a form of foraging, I think, in that I'm harnessing and working with wild yeast and bacteria for fermenting, and in this, I'm experiencing the same sense of liberation.

DC: So perhaps there is an inverse relationship between income and a healthy food culture. The more money we have, the more we are likely to "outsource" our meals to the food industry. The work of revitalizing food culture may involve disrupting the corporate/industrial regimes that encroach on our tables. Just as I realized that there was no one to pay for the oysters we collected, the experience of liberation from the monetary system that we live under, even for a brief meal, engenders a fresh perspective on our economic system. Appropriate payment comes in the form of gratitude, an appreciation for the land's abundance. A foraging experience is a chance to look at the market from the marsh, and perhaps for an instant see the strange creatures we've become. Hopefully in the process, this kind of foraging education might afford a glimpse into a mode of consciousness where we "experi-ence [ourselves] as owning nothing, as receiving existence itself and life and con-sciousness as an unmerited gift from the universe, as having exuberant delight and unending gratitude as [our] first obligation" (Berry 2006, p. 118).

HB: Our being as an unmerited gift! That's to go beyond calculative thinking (Heidegger 1971) and instrumental values (Bai 2004), in which the current world has drowned. To be able to see ourselves as unmerited gifts from the universe is to commune in unconditional love. Father Berry and his colleague Brian Swimme, the evolutionary cosmologist, present a truly breath-taking vision of our universe: that the universe is a "communion of subjects rather than a collection of objects" (Swimme and Berry 1992, p. 243). I say, foraging fits into such universe. Foraging can be an experience in communion, if we engage in it as a sacramental act of exchanging breaths and life, words of gratitude and energy of love.

DC: Absolutely. *Communion* aptly sums up what we've discussed here – the somatic, sacramental, cultural and ecological dimensions of foraging. Thank you for the delightful conversation. Be warned that I will one day invite myself to your dinner table and sample that delicious pigweed and shepherd purse.

HB: I look forward to dining with you, David! First, we will go foraging. I have to show you how to identify pigweed and shepherd purse. Then we will bring them home to prepare them my mother's way. We will make miso soup with shepherd purse and a garlic and soy sauce seasoned pigweed dish to eat with brown rice. Oh, here I am, salivating already!

References

Bai, H. (2004). The three I's for ethics as an everyday activity: Integration, intrinsic valuing, and intersubjectivity. *Canadian Journal of Environmental Education, 9*, 51–64.

Bai, H., Elza, D., Kovacs, P., & Romanycia, S. (2010). Re-searching and re-storying the complex and complicated relationship of *biophilia* and *bibliophilia*. *Environmental Education Research, 16*, 351–365. https://doi.org/10.1080/13504621003613053.

Berry, T. (2006). In M. E. Tucker (Ed.), *Evening thoughts: Reflecting on earth as sacred community* (1st ed.). San Francisco: Sierra Club Books. https://doi.org/10.1111/j.1467-9418.2008.00403_1.x.

Bourdieu, P. (1990). *The logic of practice*. Cambridge: Polity.

Dogen, & Uchiyama. (1983). *From the Zen kitchen to enlightenment: Refining your life* (T. Wright, Trans.). New York: Weatherhill. https://doi.org/10.2307/2384655.

Greenwood, D. (2013). A critical theory of place-conscious education. In R. B. Stevenson, M. Brody, J. Dillon, & A. E. J. Wals (Eds.), *International handbook of research on environmental education* (pp. 93–100). New York: Routledge. https://doi.org/10.4324/9780203813331.ch9.

Gruenewald, D. A. (2003). The best of both worlds: A critical pedagogy of place. *Educational Researcher, 32*, 3–12. https://doi.org/10.3102/0013189x032004003.

Heidegger, M. (1971). *Poetry, language, thought* (A. Hofstadter, Trans.). New York: Harper & Row Publishers.

Krakauer, J. (1997). *Into the wild* (1st ed.). New York: Anchor.

Manulani, A.-M. (2008). Indigenous and authentic: Native Hawaiian epistemology and the triangulation of meaning. In L. Smith, N. Denzin, & Y. Lincoln (Eds.), *Handbook of critical and indigenous methodologies* (pp. 217–232). London: Sage. https://doi.org/10.4135/9781483385686.n11.

Milton, J. (2012). Paradise lost. In S. Greenblatt (Ed.), *Norton anthology of English literature* (9th ed.). New York: W.W. Norton & Company.

Nakagawa, Y. (2000). *Education for awakening: An eastern approach to holistic education*. Brandon: Foundation for Educational Renewal.

Pollan, M. (2006). *The Omnivore's Dilemma: A natural history of four meals* (1st ed.). New York: Penguin Press.

Snyder, G. (2007). *Writers and the war against nature*. Retrieved 1 August 2015, from http://www.lionsroar.com/writers-and-the-war-against-nature/

Snyder, G. (2010). *The practice of the wild*. Berkeley: Counterpoint.

Stephens, D. W., Brown, J. S., & Ydenberg, R. C. (Eds.). (2008). *Foraging: Behavior and ecology.* Chicago: University of Chicago Press. https://doi.org/10.7208/chicago/9780226772653.001.0001.

Swimme, B., & Berry, T. (1992). *The universe story: From the primordial flaring forth to the Ecozoic Era-The celebration of the unfolding of the cosmos.* San Francisco: Harper San Francisco.

Thompson, W. I. (1996). *Coming into being: Artifacts and texts in the evolution of consciousness* (1st ed.). New York: St. Martin's Press.

Wilson, E. O. (1984). *Biophilia: The human bond with other species.* Cambridge, MA: Harvard University Press. https://doi.org/10.2307/j.ctvk12s6h.

Witbeck, C. (1983). A different reality: Feminist ontology. In C. C. Gould (Ed.), *Beyond domination* (pp. 64–88). Totowa: Rowman and Allenheld.

David Chang is a teacher educator in the Faculty of Education at Simon Fraser University in Canada. David taught secondary English for a decade before working as a Faculty Associate with Professional Programs at SFU. He is a Ph.D candidate in philosophy of education, currently studying ecological ethics, sustainable communities, contemplative practices and ecological ways of living.

Heesoon Bai is Professor of Philosophy of Education in the Faculty of Education at Simon Fraser University in Canada. She researches and writes in the intersections of ethics, ecological worldviews, contemplative ways, and Asian philosophies. She is also a practicing psychotherapist. Professor Bai's published works can be accessed at http://summit.sfu.ca/collection/204. Her faculty profile at SFU can be found at http://sfu.ca/education/faculty-profiles/hbai.html.

Chapter 6
Teachings from the Land of my Ancestors: Knowing Places as a Gatherer, Hunter, Fisher and Ecologist

Sammy L. Matsaw

The land of my ancestors has shaped my mind. From my mother's tribe, the Oglala Lakota who inhabit wide-open plains, I have inherited an ability to think broadly. From the Shoshone-Bannock on my father's side – peoples who live among river carved mountains – I have inherited an ability to explore the depth of thought. I often find myself crossing between different ways of knowing. My indigenous cultural understanding combined with my commitment to scientific research give me a unique set of perspectives through which I approach the world, simultaneously as a Sundancer (a sacred pipe carrier) and a scientist (Fig. 6.1).

> *According to our belief, the Buffalo Woman who brought us the peace pipe, which is at the center of our religion, was a beautiful maiden, and after she had taught our tribes how to worship with the pipe, she changed herself into a white buffalo calf. So, the buffalo is very sacred to us. You can't understand about nature, about the feeling we have toward it, unless you understand how close we were to the buffalo. That animal was almost like a part of ourselves, part of our souls.* –John Fire Lame Deer (Fire and Erdoes 1973, p. 119)

Symbolism and storytelling are the primary ways humans, as a species, explain the natural world and how we exist within it. Symbolically, the buffalo was central to a way of living and connected to a larger landscape of traditional or first foods (foods prior to colonialism), medicines and inter-generational knowledges (teachings). I come from indigenous roots, and my family culture tends to understand from an indigenous perspective, but we also live in a dominant American society. So, we've learned to live between two cultures. Storytelling from our ancestors informs us of our social contracts with our living world. In that contract: "… *all of nature is in us, all of us is in nature*" (Fire and Erdoes 1973, p. 128). For my family and me, the Medicine Wheel serves as a heuristic to perceive complex issues holistically. By allowing for this holistic perception of an issue, there is space to approach

S. L. Matsaw (✉)
Shoshone-Bannock tribal member & Oglala Lakota, Fort Hall Indian Reservation, ID, USA
e-mail: matssamm@isu.edu

© Springer Nature Switzerland AG 2020   73
J. B. Pontius et al. (eds.), *Place-based Learning for the Plate*, Environmental
Discourses in Science Education 6,
https://doi.org/10.1007/978-3-030-42814-3_6

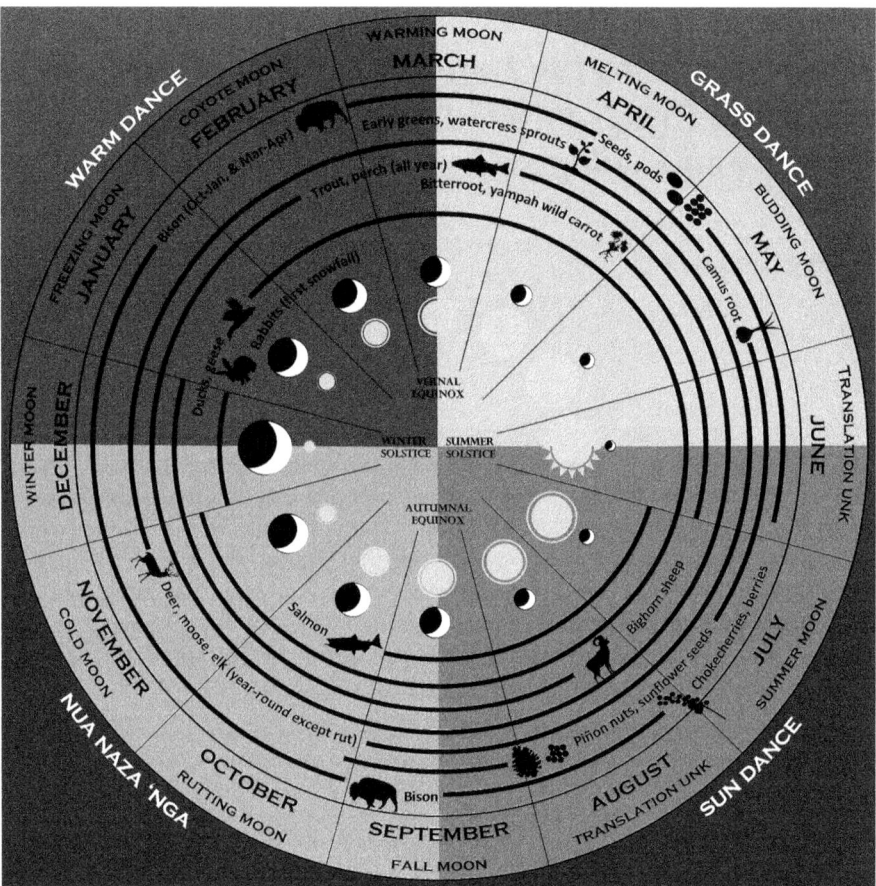

Fig. 6.1 The hunting and gathering calendar of the Shoshone-Bannocks and the ceremonies complementing the seasons by moon phase and equinox and solstice of the sun. Note: the calendar is not an exhaustive list of foods rather a depiction of the most prevalent foods hunted and gathered (recreated from Drusilla Gould, Shoshone Language Instructor)

it from a posture of humility because the issue is usually much larger than we are. This is similar to holistic ecological thinking and deep ecology minus the humility. The Medicine Wheel is a symbol embodied largely by hunting and gathering societies, and similarly, by other cultures and peoples in other forms such as the Kultrun of the Mapuche, and the Koru of the Maori.

One larger meaning of the Medicine Wheel is to find the relatedness of oneself through inter and intra relationships in the physical, mental, emotional and spiritual aspects of reality. The color arrangement and usage of the symbol varies from tribe to tribe; for instance, the Shoshones begin its ordinary and/or meditational usage facing the east, and Lakotas the west. I intend this artistic piece to be an interpretive invitation to the reader. It is meant to act as a symbol connecting a larger landscape

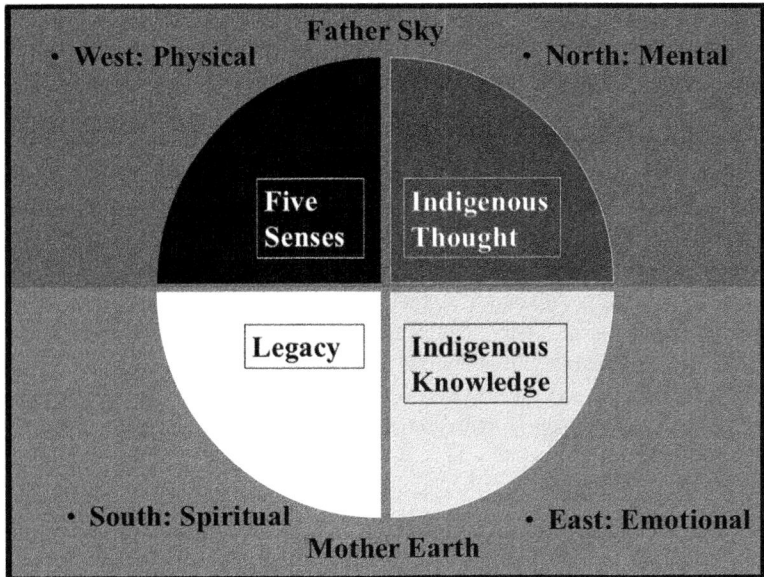

Fig. 6.2 The Medicine Wheel depiction in the inter and intra relationships with each of the quadrants respectively represented by the four directions (West, North, East, South), colors (black, red, yellow, white), states of being (physical, mental, emotional, & spiritual), and perceptions (through five senses, Indigenous Thought, Indigenous Knowledge & legacy). A propensity to see the world in this manner is also to acknowledge an embodiment of both spirits in Father Sky (as physical-mental) and Mother Earth (as emotional-spiritual)

through my ancestors who managed and passed intergenerational knowledge by eating and communing from their homelands, much like the symbolism gained from the buffalo on the Great Plains and the salmon in the Pacific Northwest. I would also suggest that readers look back at the Medicine Wheel as they read to see what meanings emerge (Fig. 6.2).

Gathering and Grandma's Place

They saw themselves as existing in a web of highly interrelated and interdependent "substances": air, water, other beings, and land. They maintained their life force by ingesting the life force of other beings. No less respect was due a wild onion than a deer. "Eat it," my father would say to us, "we took its life that we might continue our own." Eating was a holy sacrament; a thanksgiving to the creatures that provided us life. (Viola Cordova 2001, p. 4)

When I was five years old, we lived in a small, dusty trailer court across from a rock quarry just outside Rapid City in the foothills of *Paha Sapa* – the Black Hills. Even in my childhood the Black Hills invited tourists to many attractions around Mount Rushmore and the Crazy Horse monument like Reptile Gardens, Flintstone Bedrock

City and others. As a kid, I enjoyed these places, although I heard my parents talking about Mount Rushmore as a great insult to such a sacred place. I spent most of my summer with my brother and sister, and mom's younger brother and sister, my uncles and auntie playing hide and seek or just visiting by a stream under a railroad bridge. But what I remember most vividly from those hot days is gathering chokecherries with my mom's family. Later in the summer we gathered Buffaloberries (*Shepherdia*) and chokecherries (*Prunis virginiana*) along Boxelder Creek and up higher along the bench of the streambed. As we foraged, juice dripped from our chins as we freely ate the delicious Buffaloberries. Chokecherries, which have a bitter flavor and more of an acquired taste, were easier to collect without eating.

At home, my grandmother cleaned them up for making wozapi (whoa-shzah-pee, Lakota); a traditional chokecherry porridge. As the pot boiled on the stove, I would check it frequently. It smelled so yummy. How could chokecherries be so delectable in wozapi but bitter off the bush? Despite my initial impression of chokecherries, what my grandmother had going on the stove was inviting. Grandma mashed them with a potato masher as they boiled, and the rising steam filled the house with a sweet, thick cherry aroma. With a wooden spoon, Grandma mixed a scoop of government-issued commodity flour into some water, and then added it into the pot. The white slurry swirled into the deep cherry reduction creating a creamy, pastel purple pudding. She spooned some into bowls for us. I was perplexed; the chokecherries had taken on a new form, and to my surprise it was delicious! To some, "grandma's house" conjures memories of cherry pie, but for me, "grandmother's house" means thick, sweet, magenta wozapi!

As a young boy traveling into my mother's homelands, I was engulfed by the sky, and vast land seemingly without boundaries, coming into a complex web of relatives and ingesting generations of life force. Although my ancestors' times of freely hunting the sacred buffalo-*tatanka* (Lakota, *Bison bison*) on the Great Plains had long passed, I was greeted by a transformation of bitter chokecherries into wozapi. Today the Great Plains continue to teach me that some elements of life that seem bitter at first can become transformed into something else with careful nurturing and patience.

Gathering with My Daughters

I am a gatherer from a long matrilineal lineage. As I gather, I hear the stories and laughter of my aunties, mother and sister, and more recently my wife and daughters. Our early relationship with the land is to call the earth our Mother, Unci Maka in *Lakota* or Bia Sogope in *Shoshone*.

I feel that my intuition has been inherited from my mother and maternal and paternal grandmothers. As we gather huckleberries on our way home from spearing salmon in the South Fork of the Salmon River, there are a couple of places we like to stop. We use whatever containers are in our vehicle and just get to picking. Once we begin picking, several senses are aroused – I *smell* sweet fruit, woody-ness,

Fig. 6.3 My youngest son, Otaktay, and daughter, Abrianna, gathering huckleberries, an offering from our Bia Sogope/Unci Maka in August and a treat awaits our next morning breakfast: cornflakes and huckleberries, easy and delicious!

composting earth, rain-like scents. I *taste* sweet, sour, bitter, and tangy fruits. I *see* ripe berries, not-so ripe berries, medium and smaller unripe berries. I *feel* cool breezes, warmth of the sun on my skin and an intuitive sense of comfort and safety. I *notice* rock outcroppings or logs and other plants growing where berries are bigger, sweeter and tastier, reinforcing how to read the land. Mother Earth, she speaks to me. As with the many relationships nurtured in an extended family, gathering reminds me that I am connected to a much larger community of life. It reminds me to believe that traditional and medicinal foods gathering is a layer of co-health, as in, my health is your health, your health is my health, from individuals to family to community to society and Mother Earth (land) (Fig. 6.3).

Elk Blanket

Hunting rituals are performed before, during and after traditional Native hunting to acknowledge the transformation of the deer's life, spirit, and flesh into that of the human. (Gregory Cajete 2004, p. 55)

I was six years old riding with my cousins in the back of my uncle's pickup truck. It was cooler than summer with chilly nights and I was wearing tattered jeans, tennis

Fig. 6.4 Much gratitude for this deer's life and we continue to honor it by preparing it well as seen here: braised deer ribs, seared corn and onions, mashed potatoes with a deer-mushroom gravy

shoes, and a t-shirt. We were returning from South Dakota and joining my dad's large family for a hunt in Island Park, just west of Yellowstone National Park in Idaho. I was already feeling cooler from the hot noon temperature when we spotted a small herd of badeheya' (Shoshone), elk (*Cervus canadensis*) late in the afternoon. Sitting between my cousins just behind the cab, the truck came to a quick stop, dust was everywhere, and suddenly some yelling and rifle fire. As I rose from the commotion, I saw that my dad and uncles had killed two elk – a cow and a young bull. They were dead by the time I ran up to them. The men were happy, and spent some time in silence acknowledging the elks' deaths. Then we pulled the elk around to gut them. This was my first memory of experiencing the practice of our treaty rights and of manhood. I was experiencing death and life in the wealth of security found in being a provider. Carrying gut parts back to the truck, I felt proud to be with my uncles and father as they loaded the elk. On our cold, dark ride home in the bed of the truck, I laid on the elks' still warm bodies and fell asleep. When I awoke, it was to my father's proud smile as he lifted me out of the truck bed to take me inside. I was welcomed by life (reunion with my father's family) and death (hunting and killing elk) and life (feasting in celebration of reunion in Shoshone lands, through the elk's body we are welcomed home) as we ate elk meat that night back home (Fig. 6.4).

Hunting as Providing

I am a hunter from a long patrilineal lineage. Hunting has become more important in my life as a father and husband, because when I bring home healthy meat for my family, I feel as though I'm providing the best I can. As I hunt there are many practices that lend itself to mental and spiritual clarity and well-being.

Fig. 6.5 A Shoshone-Bannock Tribal buffalo hunt in April near Jackson, Wyoming. Our hunters find five nice animals who gave their lives for tribal ceremonies and gatherings

Fasting is a practice we learn in ceremony and has many applications throughout the year. As a hunter, fasting can enhance the awareness of physical senses. *Smell* and *taste* become much more acute; thus, animals nearby in the air more perceptively pass through my nose and mouth. *Hearing* focuses on the sounds made from both hunter and hunted and discerning the two. *Sight* is heightened, noticing movement, color differences, shapes, and depth. *Touch* and *feeling* is tuned to changes in temperature and direction in the air throughout the day. Looming hunger guides the direction and careful decision-making needed to find and close in on the animal. Being a provider, as with being a husband and father, makes hunting a deeply meaningful practice in my life, the connection between us to plants and animals, to clean land, water, and air (Fig. 6.5).

Hunting Salmon

Hunting the salmon is a significant part of our way of life. The name for the salmon, Agai, has been used to define our people as the Agaidika [salmon-eaters]. No one can understate the importance of this resource to the Shoshone and Bannock peoples. We have continued to exercise our right to hunt salmon in the Columbia River Basin since the Treaty was signed. The Shoshone-Bannock Tribes are today co-managers of the anadromous fish resource in the Columbia River Basin and have continued to work towards improving the habitat and supplementation efforts. (Lionel Boyer, then Chairman of the Shoshone-Bannock Tribes 2000)

"Coming to know" processes (research processes) and the role salmon play as a significant part of our way of life began a few summers later from my first fall hunt in Island Park. We had taken a road trip to the Yankee Fork of the Salmon River, which originates in the Salmon River Mountains just east of Stanley, Idaho. We were hunting *agai* (Shoshone), Chinook salmon (*Oncorhynchus tshawytscha*) in knee-deep water with spear poles about 12–14 feet long. It was a hot July day and we had been walking upstream all day. Salmon were driven almost to extinction in the Yankee Fork by dredge-mining from 1940 to 1952. There were so few salmon that day, we were searching every part of the stream and the men were using the butt of the spear to flush anything out from under the banks. They were also scoping. A scope is a tube about 3 feet long with clear glass at the end. It works to see into the water like a periscope for a submarine but in reverse.

There used to be so many salmon in the streams, if you tried to walk across they would trip you. My mom, siblings, and I were in some disbelief of these stories they told. How could there be fish so big in these small streams? As the day went on it didn't seem believable there were any fish in the stream at all, much less fish so big they could knock you over. We stopped at a deep hole when my dad, with his spear, looked over my uncle's shoulder as he was scoping. He signaled with his hand and pointed. My dad took the scope and looked, smiled, and gave it back. He handed his spear to my uncle, who gently and carefully entered the spear into the stream and lined my dad up. With a cigarette in his mouth and both hands on his spear, while my uncle, with one hand on the scope and the other on the spear gave a signal to my dad, who then speared! The salmon was on and pulled my dad into the creek head-first. Again, I was perplexed; my dad had just gotten pulled into the stream by a monster. The story was true! He came out of the water fighting the salmon onto the bank. It was massive! My uncles and dad jumped on it as it flopped and flipped. My dad took out his Old Timer pocket knife and stuck it in the head; they fell silent and paused to acknowledge the salmon dying. The tail fluttered with a couple last slaps on the stream bank. The men were happy again, and so was I.

Traveling from my mother's homelands and back to my father's has shaped a wholeness from each part of my parents. The engulfing sky of the Great Plains expand my thoughts, and the river churning waters carving the mountains along canyon walls speak to me, focusing my thoughts. Neither of these ways of thinking/knowing are exclusive of who I am nor how I think/know. During the summers when I return with my wife and children to Oglala country, we've spent time learning how to set up a tipi Lakota style with my relatives. Inside the tipi, I've come to know where my ancestors would find concentrated thought, a place to bring expansive thought into focus on wide-open plains. And just as well in Shoshone-Bannock country I've been traveling ridgelines hunting and have looked far across the Snake River Plain expanding my focused experiences up high from down along the river below. Between either my father's country or my mother's I can find the place for my thoughts and understanding to open to possibilities of the sky and focus clearly like water.

Long Spears

I've come to think and interpret the world through my inherited intuition and intellect of my parents: I am a spearfisher from a long lineage of indigenous peoples of Turtle Island. I've thought the act of spearfishing on the Salmon River, at times, arouses feelings of how it was to spear buffalo from horseback on the Great Plains. As I spearfish, there are practices and senses aroused. I stand holding my spear listening to women and children yelling on the banks above, "Coming Up!" and "Going Down!" as they can see the salmon moving up and down the stream. The spearfishers are moving, stationing themselves on large rocks; our positions complement one another strategically. I notice that where I'm standing, a smaller stream enters. We are surrounding the tributary entry into the main stem, and behind me there are huckleberry bushes.

Tributary junctions are special places where two streams come together, and where salmon smell the water and know which direction to go. They return from the Pacific Ocean, over 900 miles in the Upper Salmon River, to find where they were born. Salmon are born in the gravels of the streams high in the Rocky Mountains, then move to the Pacific Ocean to gain 95% of their body mass and return back into freshwater on an amazing journey (in most cases this is terminal). At tributary junctions, Chinook salmon move back and forth, up and down the stream, making them vulnerable to spearfishing. Indigenous knowledge of these places and how to successfully hunt salmon has been passed from generation to generation, bringing with it place names, stories and cultural links, connecting me back to my ancestors. Long-held family wisdom brings me to this place where my ancestors stood many times before me.

The huckleberry bushes behind me are growing where the seeds were transported in the guts of my ancestors from gathering berries as the harvest seasons are close to one another. We were among other animals dispersing ocean nutrients from eating salmon and seeds from eating huckleberries. Together these are ripe conditions for the successful growth of huckleberry bushes in the future as they are left on the ground from passing through our guts. Thoughts of places long ago enmesh with clarity about where I stand today and draw me to the exact same sites where my ancestors once stood. I imagine my descendants gathering huckleberries and salmon that we will leave for them now, to be realized after we come to pass (Fig. 6.6).

Looking Forward

In this chapter, I am sharing my stories, honestly, with diverse readers who share experiences on the same landscape embedded in a continuing tragic history. Through boarding schools, urban relocation programs and policies to end our culture, our peoples were in varying degrees displaced physically, mentally, emotionally, and spiritually from the lands that once sustained them. Generations before me, my

Fig. 6.6 Salmon giving their lives so that we may go on from the South Fork of the Salmon River in mid-July, a good return of fish to the basin for the summer (myself on the left and Nikolas, my wife's brother)

parents, grandparents, and their generations have been through so much, yet they still shared, raised my siblings and cousins as best they could. We were categorically, by US standards, in poverty but lived such a rich life connected to water, land, and our traditional/first foods. I came back to my own deeper experiences with ceremony, culture, and community when I returned to the homelands of my parents from a stint in Iraq (2004–2005). Through my own healing processes, I am unpacking normalized concepts of toxic masculinity, internal and external oppression, survivor's guilt, and so on to see empowerment, resistance, and freedom in direct connections to land with my communities. What I see in myself is a reflection of what I see in my communities and my communities sees themselves through me, we are one.

I want to stress the importance of bringing family along. With our families on the land we are setting out to disrupt the cycles of toxic masculinity that plague American society. As part of a matrilineal society we respect women as leaders. They open the seasons, the grounds, the taking of life because they bear life as the nation-builders and have that responsibility, only then are men able to hunt, fish, and gather first foods. The ceremonies and protocols are asking for permission, a consensual engagement that must be renewed, and renewal is ongoing. For instance, for the Shoshone-Bannocks the salmon season begins with a sweatlodge ceremony in the headwaters of the Middle Fork Salmon River in Bear Valley. The sweatlodge is a representation of a woman's womb, and the ceremony is a process of rebirth, an acknowledgment of life bestowed by the life we ingest from the womb of our mothers. For the Oglala Lakota, at the center of the Sundance grounds we stand a cottonwood tree that was taken from a distance away and facilitated by the blessing of young women who prayed allowing us to cut it down. Before we plant it back into the ground those same young women pray with water, select parts of the buffalo, and chokecherries as a ceremonial offering to the More-Than-Human (MTH) sacrifice so that our lives may go on. After the prayers they add them into the hole to feed our tree while the Sundancers take on the next four days without food and water. The complete, complex, consensual commitment through our culture, customs and protocols involve roles through each of our family and relatives towards our children to perpetuate ceremonies honoring water, land and MTH life beings. Our women, children and elders bring other perspectives – when we see the world through their eyes, we become better hunters, gatherers, fishers, and human beings. Bring your families and acknowledge your relatives. Through these acts they, us and we are disrupting ideologies of me, mine and I.

I'm talking from an inclusive we, us and ours as in Indigenous waters, lands, plants, animals, and human beings. Although this in most ways excludes a larger part of American society, it also challenges settlers to be better neighbors to Indigenous life: living and non-living entities. To go deeper into colonialism, ideologies, and methodologies. To ask what is decolonization? Does it benefit me to decolonize? To truly sever ties of imperialism as promised in 1776. I can say 'we' need that more than you can know.

I also want to emphasize the importance of keeping up our relationships with ancestral waterways and openness to non-native people in some of our experiences. Each summer, with the help of my white colleagues, their families and ours, we journey with young Native peoples down ancestral rivers. We are attempting to connect pristine river corridors with Indigenous Knowledge in its complexity. It's significant to have young people thinking and knowing these places as did their ancestors and being able to return to places of cultural, ecological, and educational importance. The inclusion of non-Tribal members is important to understand our way of life as truly invested allies through a first-hand experience. Much of our culture has been lost because non-Native folks didn't understand, or didn't want to understand who we are, and how we were living on the lands they wanted. Our removal and displacement was intentional, deliberate and costly. However, trauma goes both ways. For us to come back from that, non-Native folks will need to come to understand us, who we are and how we are re-initiating living on our homelands off-reservation (Rez) through first foods. Our time on the river has been priceless and healing in multiple ways and has begun to reduce the distances of cultural divides.

We have discussed how doing 'research' as a coming to know process draws on parallels from hunting, gathering and fishing with the land, people, plants and animals. Because we bring non-Native folks, sanctioning research through ceremony as an act of consensual engagement has been challenging and rewarding. One experience in particular during a seven-day trip on the Middle Fork Salmon River provides a glimpse of possibilities, a renewed way of comingling with a landscape using lenses of IK and Western STEM methods can engender profound learning. The setting was at this beautiful cultural site called Veil Falls, a natural cliff amphi-theater with a water fall misting over the middle of it. Our friend, a Cherokee Citizen and a snail biologist, had a certain interest in describing a terrestrial species of snail (*Oreohelix*). He had found nothing so far, and this was day 6 of 7. Previous to the trip, he had been looking at maps, studying the geology to look for marble outcrop-pings or other sources of calcium that snails need for their shell building. Once on the trip, day in and day out, he was searching along the river. We selected camps or stops to hike up to marble or limestone deposits and had no luck finding a viable snail presence.

My wife initiated a Chanupa (pipe) ceremony after a bit of a hike up to Veil Falls. The youth, a high school teacher from the Rez, undergrads, grad students were also conducting ceremony, through songs, and dancing with the waterfall. At this point our Native land snail biologist had all but given up on the snail searching and was enjoying the river, the beautiful waterfall overhead and soaking up some sun on a huge rock. Just behind him a youth found a snail shell and asked him, "is this a snail shell?" He grabbed it and looked closely, "Yes! Where did you find it!?" From then on, he was in full snail biologist mode. He had a few kids and adults enact a search protocol (hunting/gathering) for more snails. We found two species; *Oreohelix* was among them. This happened while the ceremony was going on, because of cere-mony, and as part of ceremony. Through ceremony similar to those around hunting, gathering and fishing, we were asking for permission to engage with the land in both ancient and contemporary, Indigenous and Western ways of knowing. By honoring our relationship to the land, we were able to open up our minds, bodies, and senses to what was before us, and in so doing we gained insight into the old as well as contributing to the future care of this place. May the land continue to teach us and show us how to care.

In closing, if caring and consent begin to shape characteristics of American iden-tity, perhaps the sense of feeling lost will subside. Speaking in classrooms of mostly white students I've heard their reflections and questions with candid notions towards a sense of homelessness. In my story are guides using symbols such as the Medicine Wheel and what first foods to hunt, fish and gather during our seasonal round calen-dar. I'm sharing acts of my family as they have taught me how to care about our homelands and ask for consent to take MTH life. Today, caring for clean water and healthy food sources is an everyday act of resistance. Consensual agreements of cooperation between MTH life forms is how we survive(d). These reciprocating acts of being a human are our contracts to land, water and first foods for us to take care of them and for them to take care of us. Through these two acts of caring and consent, among many other teachings, we nurture a sacred relationship with place, and this is how we feel at home. From Indigenous Ecology I'm saying to study home (oikos-home, ology-study of, ecology) and be Indigenous can bring forth new ways of being a human and an ecologist who understands two cultures and connections to land through first foods. My hope is to make ways forward for more people like me and, variations and complexities of who we are.

Suggested Readings

Bang, M., & Medin, D. (2010). Cultural processes in science education: Supporting the navigation of multiple epistemologies. *Science Education, 94*, 1008–1026. https://doi.org/10.1002/sce.20392.
McGregor, D. (2004). Coming full circle: Indigenous knowledge, environment, and our future. *American Indian Quarterly, 28*, 385–410. https://doi.org/10.1353/aiq.2004.0101.
Simpson, L. B. (2014). Land as pedagogy: Nishnaabeg intelligence and rebellious transformation. *Decolonization: Indigeneity, Education & Society, 3*(3), 1–25.

References

Cajete, G. (2004). Philosophy of native science. In A. Waters (Ed.), *American Indian thought* (pp. 45–57). Malden: Blackwell Publishing.
Cordova, V. F. (2001). Time, culture, and self. *APA Newsletters, 1*, 3–5.
Fire, J., & Erdoes, R. (1973). *Lame Deer, seeker of visions*. New York: Simon and Schuster.
Lionel, B. (2000). *Hearing, Senate. Columbia river power system: Biological opinion and the draft basinwide salmon recovery strategy*. Washington: U.S. Government Printing Office.

Sammy L. Matsaw holds a PhD in Water Resources. An interdisciplinary dissertation on native freshwater mussels led him to understand that his own success was by respectively interweaving Indigenous Thought and Knowledge with Western Science and Thought through symbology to inform both life ways. In his lifework, he introduces tribal youth to the idea of bringing Indigenous culture into Western pedagogy. He and his wife are creating an inter-cultural STEAM (Science, Technology, Engineering/Education, Art, & Math) pedagogy more agreeable with Indigenous peoples.

Chapter 7
Catch of the Day

Alison Laurie Neilson and Rita São Marcos

May 2015 – We wanted to go to the fish market before it closed in the morning, but we were working so late the night before… the man who sold us the fish at the supermarket in Ponta Delgada, São Miguel Island said we were destined for a great feast as it was fresh from the local waters, and the owner of our rented house was excitedly taking over the preparations following a traditional Micaelenses recipe to impress his childhood friend, "The Great Baptista", a local fisher and activist who we had invited to dinner. "Definitely not caught in the Azores" Lurdes tells us, after examining the flesh of the fish – the first time she has agreed to eat fish that her family has not caught and she herself has not cooked.

A Conversation About Eating Fish

Is it serendipity, or the force of some global ecosystem process, which brings us to this place and conversation? We find it difficult to know where to start this chapter since both of us as well as anyone reading is already in the middle of many stories about fish, fishing and education, and because, as Thomas King (2003) reminds us, how we tell a story is important: it can kill or cure. Many environmental problems in particular are "produced, reproduced and intensified… by the ways in which we live our 'storied lives'" (Gough 1993); so, perhaps the best we can do, is to start with ourselves.

A. L. Neilson (✉) · R. São Marcos
Centre for Social Studies, Colégio de S. Jerónimo, University of Coimbra, Coimbra, Portugal
e-mail: aneilson@ces.uc.pt; ritamarcos@ces.uc.pt

© Springer Nature Switzerland AG 2020   87
J. B. Pontius et al. (eds.), *Place-based Learning for the Plate*, Environmental
Discourses in Science Education 6,
https://doi.org/10.1007/978-3-030-42814-3_7

Alison: Growing up in Canada far from the ocean, my best fish dinner would be the occasional, thickly battered and deep-fried halibut from a "fish and chips" shop, undoubtedly frozen for the transportation after it was caught, and a favourite in my family because it had very little "fishy" taste. My family spent many summer weekends camping at lakes but only very rarely did any of us sit with a fishing rod amongst the dawn or dusk swarm of mosquitos. Later, in the mid 1990's, in the wake of the collapse of the Northern Cod, I remember watching the television news covering the "Turbot War", showing the seizure by the Canadian Coast Guard of a Spanish boat which, according to the national news, had been "killing baby fish". But it wasn't until 2008 when I found myself, living on a 17 × 24 km piece of volcanic rock, hundreds of kilometres from the coast of Portugal, in the middle of the Atlantic Ocean, that I began to learn about and have a relationship with fish and fishing communities.

> *October 2011 – Months of preparation, traveling between the nine islands to identify the issues to discuss amongst fishers, local fish merchants and scientists who work in small-scale fisheries in various parts of Europe: fisheries policy, commercialization of fishing products, management and monitoring, fishing tourism, and education and training… it has been an intense morning with some loud arguments and frustrations at times. We are all hungry when we enter the newly built cafeteria of the University of Azores overlooking the beautiful rocky shore of the Atlantic Ocean, a few kilometres from the fishing port of São Mateus on Terceira Island, over 1500 kms from the continental coast of Portugal. "Peixe?" receives immediate nods of affirmation to have fish for lunch. Nile perch is scooped onto our plates.*

Rita: My last name São Marcos is quite unusual and uncommon for Portuguese, so from an early age, my name made me realize that I come from a family of cod fishers from Ílhavo. This small city in the north coast of Portugal is the source of the national identity celebrated by the dictatorship; the hometown of the lonely dorymen of the *faina maior*, the White Fleet who fished in the Grand Banks of Newfoundland. The challenge for me and continuous struggle to have my name spelled correctly by teachers, governmental officials and others, fed my curiosity to question my father on the ancestors I actually never had the chance to know. I was not aware of what it meant to be such a heroic seaman until I heard about my father's experience as a young man in a time when Salazar was trying to maintain control of colonies in Africa. Being of an age selected for conscription, my father had the choice to join the army or to escape going to the Colonial war if he went to fish cod. He chose the war. Can you imagine? Choosing to go to war seemed safer than going to fish cod.

Fisherman, think well! God, Country, Family and the Sea ought to be for you,
Fisherman, the four cardinal points that guide your life.

Fisherman! Don't view work as punishment because it is the noblest and most honourable occupation of man.

Fisherman! In the struggle to contribute to your country, there are no lowly professions; all are equally honourable when the Love of work exists.

The greatest virtue of the worker and the soldier is discipline.

Excerpts from speeches by Salazar published in the Jornal do Pescador (Fisherman's newspaper) (Cole 1990)

Statue of Caspar Cortes Real in St. John's Newfoundland in honour of fishermen known by locals as "the iron men in the wooden ships". (Photo: Alison Neilson 2009)

Alison: I spoke with a man of a similar age of your father, who in 1964 decided that he would not go fight in Angola and Mozambique but instead chose to go fish cod: a decision, he came to realize, which was probably a mistake. Besides the horrendous work and living conditions aboard the ship, he talked about hours alone on a small dory in the fog in the north Atlantic, unable to see other dories or the main ship. He remembered weeping endlessly out of regret, loneliness and fear of leaving his family without a father as a slip into the icy waters would mean certain death. He described the dangerous maneuver of bringing the fish and the dory back onto the deck of the ship and an accident when only the quickness of a crewmate saved his life after he fell in the sea between the two vessels (interview António Soares, 2009, Toronto).

Rita: Being Portuguese, fish has always been dominant in my life. We are the people who eat the most fish in Europe (56 kg/person/year vs EU average 22 kg),[1] and there are over 1000 ways to cook bacalhau, dried cod, which is a staple in the kitchen and *fiel amigo* (faithful friend). At twelve years old, I did the unthinkable: I chose to become a vegetarian and not even eat fish. I had begun to learn about the inhumane, non-environmental and unjust ways that many animals and people who

[1] European Atlas of the Sea http://ec.europa.eu/maritimeaffairs/atlas/index_en.html

work in industrial meat and fish production are treated. I had to explain to those who did not understand why I would do this, and that it was actually possible to live and physically grow into an adult without eating meat or fish. It was tiring to explain over and over again and frustrating to receive only potatoes and salad when I ate at the homes of other people. As a young person with few supports to help me eat a fully balanced vegetarian diet, I felt that I had failed when I was forced to re-introduce fish into my diet to prevent anemia. At first it was difficult both emotionally and physically to eat fish again. A friend joked that this was "karma", payback from my fishing ancestors.

> *Although, I often look at a fish, at the fish that I caught, and I look at him, and he had to die for me to live. And that's it, life in the end is also showing this truth. So I can survive and all those who eat fish, it is necessary that the fish die. That's it, and we are all part of this chain. So, the biggest eat the little ones* (Sr. Genuíno Madruga, Horta, Faial Island, Azores Portugal, 2009 interview).[2]

Alison: It can be a frustrating when we try to embody our beliefs, when we are regularly taught to undertake ethics at an individual level, not at a collective or political level. When I introduce myself as working on environment education issues with fishing communities, I often hear the rhetorical response: *so you are teaching the fishers to not over-fish, eh?* Sometimes, when I explain that not all fishing and fishers are alike, that there is a difference between industrial and artisanal fishing, people even warn about romanticising the coastal fishers. Having done extensive ethnographic work with fishing communities, while living in the islands, I acknowledge the diverse mix of interests, experiences and histories and, most importantly, the ability and the right of fishers to define themselves and their traditions as culture changes and adapts via intercultural encounters. In a conference of Azorean fishers with local and visiting foreign scientists about policy issues, fishers spoke of feeling excluded from discussions by top-down processes at local and European Union levels (Bulhão Pato et al. 2011). So, I invite people to learn about the life of fishers, and listen to what they are saying, to listen for the underlying understandings of the world, the understandings that are based on experiencing the world directly, the praxis of living in tune with the ecosystem. For instance, when Kevin from St. Bride's in Newfoundland told me about getting sick every single time that he goes out to sea, but that he has not considered doing any other "work" for the more than 30 years of his adult life, I marvel at the power of the ocean and fish to keep him there. His wife said that he always comes back from the sea with a great weight loss. But Kevin says that he loves it, so as someone who gets car sick regularly, I want to dig deeper into what he is saying.

---

[2] "Embora, muitas vezes eu olhe para um peixe, para aquele peixe que pesco e olho para ele, e ele teve que morrer para que eu viva. E é isto, a vida no fundo também passa por esta verdade. Para que eu consiga sobreviver mais todos aqueles que comem peixe, é preciso que o peixe morra também. Pronto, e todos nós fazemos parte desta cadeia. Enfim, os maiores comem os mais pequenos." (Sr. Madruga)

Rita: With the strong images of industrial fishing that dominate; the individual fisher is mostly invisible in the media. Which is why it is surprising to find out that more than 90% of the total fishing people in the world fish at a small-scale, and that this type of fishing exists in rich, technologically developed places of Europe and North America as well as in other "poorer" places (Rocklin 2016). It is quite amazing how robust the European fishing communities are in the midst of all the economic and social upheavals. Even with the privatization of fishing rights, the globalisation of the trading in fish and the regulation of fishing, there are still 90,000 full and part-time jobs in small-scale fishing in the EU (Symes et al. 2015).

> *For me the sea is life. The sea brings life. If you don't respect it, it will take your life. But if you respect it, you will always find bounty from the sea, there's always reward in that you know, you learn from the ocean, you learn about the animals that live and survive in the ocean, their ability to adapt, to change, between water temperatures and wave and sea conditions, it's probably the most unique spot we've got on the planet, it's the biggest volume we have of mass on the whole planet, and for the human race, we know more about outer space than we do about the deep ocean ... and the more we learn, the more we see, the more we want to know, the more we'll go hunting for more knowledge, and the ocean has all the power. It's all powerful. If you don't be careful, the ocean will consume you pretty quick.* (Joseph O'Brien, Bay Bulls, Newfoundland, 2009 interview)

Alison: Governments often provide training for alternative livelihoods to fishing, based on the strongly held belief that fishers are poor, ignorant and it is inevitable that their livelihood will die; therefore, they need to be taught other work (Jentoft et al. 2010). As a young biologist with a dream of saving the environment from selfish people, I did not question such a perspective. But to assume that fishers are unskilled and lack knowledge ignores that running a boat in the ocean requires skills in fixing engines, repairing wood and other materials, flexible decision-making regarding reading the weather, the location of the fish, understanding changing regulations requiring financial investments in gear, fishing quota and a myriad of dynamic complex processes of the ecosystems of which they are a part (Hind 2015). Instead of following a 9 to 5 schedule, fishers can make their own decisions as to when to go to sea; they were free instead of having "free time" outside of "work time" (Højrup 2003). The wind, the rain and the mood of the ocean controls when fishers are allowed into the sea. The boats are small and do not carry much gear and the waves on the other hand are big, especially in winter; some Iceland fishers have even suggested there is no need for governmental interference in fisheries management as nature itself regulates the fishing effort (Einarsson 1993).

Research and Policy Development: Power Struggles for Fisheries

Rita: As we learn about formal governmental procedures, regulations and research strategies, to understand how human societies can eat from the sea today and still into the future, we are offered reams of population studies of fish, quotas and the physical ocean, but little about diverse histories and strategies of conservation from fishing communities around the world. The Azorean government's strategy for fishing from 2015 to 2020 (Governo dos Açores 2015), however, surprises as it is based on the idea that fishing is about and for people; the ocean is a "peopled seascape" (Shackeroff et al. 2009). Their main approach is to enhance the income of the fishers and improve their working conditions. They are concerned with preventing overfishing as well, so they seek to have the fishers *fish less but sell better*. Dignity of the work, social security in the context of the weather and the quotas allotted to the Azorean fishers are highlighted in the governmental document, and they seek collaboration with fisher community associations to help make the strategies work. Pointing to five problems for low income, they seek specific strategies related to the nature of the economic structures in place and do not take the lazy approach of using education as a panacea for these non-education problems: refreshing given the ubiquity of "education as universal solvent" norm (Converse 1972) that we have sadly come to expect when reading government and fisheries biology documents.

Alison: On one hand, I have sympathy for educators trying to deal with complex environmental issues, as they must trust reviews of scientific literature and are not always able to engage with the continuing debates and contested ideas – the hallmark of science – for understanding all elements of all issues. On the other hand, following natural and physical scientists blindly without considering social research is a continuing frustration for me. Literature which should be trustworthy can have important gaps. A report by the European Academies' Science Advisory Council and the Joint Research Centre of the European Commission (2016), on marine sustainability betrays a woefully limited understanding of marine issues: that which ignores all social sciences (Symes and Hoefnagel 2010) and fishers' knowledges (Hind 2015), similar to my experiences in policy meetings in the Azores (Neilson et al. 2016). The working group of 20 "European science academics" and the eight peer reviewers are composed entirely of natural and physical scientists without a single social scientist (anthropology, sociology, psychology, education), nor economist. The report claims to provide evidence-based scientific support for EU policies and cites 70 references from which the evidence is drawn, and the recommendations based. Of these papers, 70% are about natural and physical sciences, and the rest are documents about existing policy. Nevertheless, the lack of specialists or social science research did not prevent the working group from making statements and recommendations about demography, economy and education. They call for further support for research without showing that the problems of marine sustainability are caused by lack of knowledge or innovation. Meanwhile the sound of Blue Growth rings through the halls of Brussels (EU Commission headquarters), and ministers of fisheries and marine affairs attentively listen to heads of banks and other financial institutions committed to shareholder profit give keynote addresses at international Ocean congresses extolling the virtues of privatization and enclosing the ocean, without hearing from the economists and social scientists who offer strong critiques to this approach of private property as essential for conservation and the Tragedy of the Commons, a long disputed myth evoked still by bankers and many fisheries biologists.

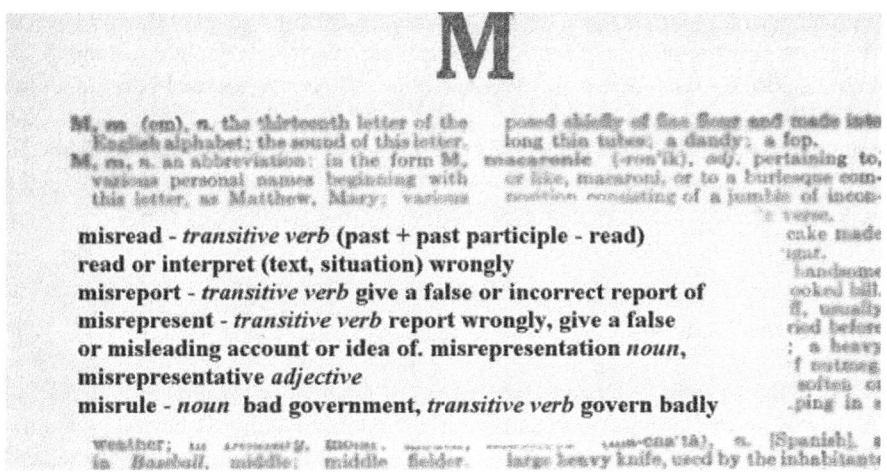

misread - *transitive verb* (past + past participle - read)
read or interpret (text, situation) wrongly
misreport - *transitive verb* give a false or incorrect report of
misrepresent - *transitive verb* report wrongly, give a false
or misleading account or idea of. **misrepresentation** *noun*,
misrepresentative *adjective*
misrule - *noun* bad government, *transitive verb* govern badly

Rita: What does this council and EU commission mean by "interdisciplinary integrative" approach when they leave out economics, social sciences and education? Continuing the global environmental discourse that humans are alien to the ocean and their presence is always harmful (King 2005), the report offers little help with the complex "wicked" (Rittel and Webber 1973) issues inherent in seeking sustainability. In reviewing the ethics behind the changing management approach to fisheries, which arguably could apply to broader marine sustainability, Carolyn Merchant (1997) called for a partnership ethic in which the needs of both fish and people are privileged, since "the greatest good for human and nonhuman communities is in their mutual living interdependence" (p. 29). She included consideration for cultural and biological diversity, relationships and obligations. Many others agree (Olsson et al. 2004) with her and suggest that the most direct way to include cultural diversity in management approaches is to include a diversity of people who have different perspectives of the sea and fisheries, particularly local fishers (Neves-Graça 2004).

Alison: Unfortunately, unexamined but powerful ideas and images underlie public and scientific opinion effectively limiting successful policy options, despite claims of rigour in the work of fisheries biologists, oceanographers and other natural or physical scientists. These images implicit in governance of fisheries need to be deliberated openly in order to clarify what is driving the goals proposed for policy. The earliest idea, that drove unlimited industrial fishing as opposed to continuing of indigenous and other small-scale fishing, was that the seas are inexhaustible, legitimized in 1885 by Thomas Huxley's ideas, followed by Garrett Hardin's (1968) perverted misunderstanding of diverse environmental history giving biologists, and environmental educators the seemingly unsinkable "fact" that as a species humans will selfishly destroy any commons which are not protected by fences and privatization (Jentoft et al. 2010). Current policies and policy makers ignore fishers' sociocultural identity and relation with the sea and see small-scale fishing as old-fashioned with no future, and upon that assumption push these fishing people to become workers in fish farms or processing and packaging units (personal communication I. Ertör Feb 01, 2017). Across sectors the privatization argument promoted by most, but not all economists serves to dictate government policy with further detrimental impact on the livelihoods of small-scale fishers, helping to promote the image of fishers as wallowing in poverty in need of rescue from family run fishing boats and free access to the ocean via 9-to-5 jobs in private industry away from the sea or working within industrial fishing companies.

> *The sea, for me, is everything. I am nothing without the sea. The sea is everything to us*[3] (Sr. Rúben Dutra, Santo António, Pico Island, Azores, Portugal, interview 2009).

Rita: Critics of the neoliberal assumptions of inherent selfishness of humans and the assertion of Hardin (1968) and the EU Common Fishing Policy, are common (Lam and Pauly 2010), yet the metaphor, Tragedy of the Commons regularly pops up in academic and public forums and without challenge even in writings that inherently question its validity (e.g., Pierce et al. 2012). Opportunities exist to know fishers as mindful stewards of the sea, but unless educators seek and share these stories the cycle of re-affirming harmful myths continue. Corvo Island, located at the westernmost point of the Azores archipelago is unique in having the only marine reserve in Portugal established at a community level: a working idea from a local tour operator and fishers who proudly and actively raise awareness and enforcement of Caneiro dos Meros (Abecasis et al. 2013). This is one such story; in coastal communities around the world, more are waiting to be heard.

Alison: Yes, that is a good example showing that fishers and biologists alike, are concerned about fish populations, but, while management schemes are contested, superficial statements about overfishing regularly go unchallenged in policy discussions in which I have been present (Neilson et al. 2012). Occurring in meetings in the Azores, it is quite surprising because these same researchers and politicians promote artisanal small-scale fishing that occurs in these islands as a sustainable practice (Carvalho et al. 2011). Furthermore, European populations continue to eat more fish than are caught in European waters, consuming fish from African and Asian waters while continuing to blame the fishers for over-exploitation of the oceans. On top of that, different species of fish exist in the ocean, but around the world, the same "White" fish are sold commercially in restaurants and shops, most of which does not contain any information on how it is produced, where, when and how they are caught. Neither is it possible to track who gains from such trade. Environmental justice issues related to distribution, recognition, participation and capabilities aspects generate socio-environmental conflicts between small-scale fisher people and fish farms (Ertör and Ortega 2015). Most of the fish sold come from aquaculture production, which has several potential and observed negative socio-ecological impacts. For instance, if the fish is carnivorous, the Fish Conversion

---

[3] "O mar para mim é tudo. Eu sem o mar não sou nada. O mar é tudo para a gente." (Sr. Dutra)

Ratio, i.e., the fish component of the feed that farmed fish has to eat in order to gain 1 kg weight, is usually higher than 1. So, in fact, we are consuming even more wild fish in order to produce less volume, but higher economically valuable farmed fish instead of eating the wild fish (Tacon and Metian 2008). Although promoted as a solution for world hunger, it is not clear to what extent aquaculture contributes to food security (Béné et al. 2016).

Rita: You shared your frustration over the reaffirmation of powerful myths, but it was not until I was at a social science conference focused on fisheries that I felt the profound weight of these unexplored assumptions. We could say over-consumption instead of overfishing to redirect the focus to our own complicity or question the damaging effects of oil exploration and drilling, marine traffic, military exercise and other exploitive activities on the sea. Yet, the overfishing story provides a simpler argument than wading through the complicated concepts and processes of central-ization, capitalization and marketization of neoliberal management systems, all pro-tected by powerful interests (Høst 2010). It is also disappointing to know that over the past decade, researchers are showing increasing reluctance to publicly critique government policies since these policies also drive the research funding opportuni-ties of which we need to survive (e.g., "unravelling of academic integrity" and "cor-ruption of academics as custodians of truth" Chubb and Watermeyer 2016).

Alison: Sometimes I question my role as researcher when I know about the need for more direct actions, such as working in policy departments in governments as I did after I studied wildlife biology or spending time in the ports to make sure that other sectors, tourism, for example, is not continuing the grossly unbalanced, and unfair blaming of fishers for the massive plastic problem in the oceans. How can I point my finger at any fisher who may throw a piece of trash in the ocean, when I am wearing a polar fleece jacket which sheds microfibres every time I wash it? An "every piece counts" argument is only fair if awareness and blame is shared propor-tionately to the level of responsibility. Military actions, regular shipping procedures such as emptying bilge water, as well as spills and other accidents – all part of enabling me to wear the polar fleece jacket, as well as governments allowing gar-bage to be dumped directly or indirectly into the sea, is massive compared to a casu-ally thrown cigarette package. Fishers are not able to stop this massive misuse of the oceans, yet they suffer the consequences of such degradation via less healthy eco-systems from which to fish, threatening their livelihoods, heritage and ability to feed themselves and their families. This inequity of benefit (small-scale artisanal fishers see little of the 25€ for a plate of Cherne in a Lisbon restaurant) and negative impact of all of those Decathlon jackets creating more and more plastic islands, is an issue of environmental justice.

Rita: Coastal fishing communities as an entrance point to other lands always have been on the forefront of new arrivals, be it plastics or people. Primarily because of location, few of the current refugees come to Portugal, and the Azores Islands are particularly far from the routes of people escaping violent conflicts. However, in other places, small-scale fishers are face-to-face with another important world crisis aside from fish. Since 2015 refugees from Syria and Afghanistan have been seeking the shores of Greece via the Aegean, Ionian and Mediterranean Seas. Apart from the direct loss of livelihood for the fishers resulting from the discarded life vests, boat remains and other garbage polluting the land and water, damaging fishing nets and driving tourism away, the fishers, especially those on the island of Lesvos, have been alone in most of the rescue efforts. Working day and night to save people, they are also exhausted with no time to fish and are suffering from the emotional toll of finding bodies in the water and washing up on the beaches. The complexity of the human tragedy also complicates the relationships between fishers in the communities who struggle with economic crisis while also trying to help the refugees who remain in their villages (Vlachopoulou 2016).

"I have entered another world, legally, but which feels part invasion, part trap, as all eyes seem to question if me and my young looking assistant are lost tourists. I sit on the cold chair trying to take in every detail of the fishing Lota, where the incoming boats sell their catch. The early morning calm of Ponta Delgada infiltrates this auction where the only sounds are those of the conveyor belt, the faint clink of the bidding devices and the occasional whisper some bidders make into cell phones, checking and clicking. Beautiful colours of glistening scales on delicate creatures and strong firm bodies of others are easy to imagine alive in the nearby water only hours before. I watch as tray after tray of fish roll past the camera as numbers swirl on the monitor: name of boat, name of fish, kilograms, price of sale. My assistant and I look to each other as we simultaneously realize that the numbers are going down, not up. I will later learn that this is normal for all fish auctions, unlike auctions for cattle, milk, gold and other valued and rare items which go up in price during the sale process. But at this moment, I feel sick and in the place of the fish in the trays of ice, I see the faces of fishers who have told me of their struggles and I see the boats that I know have been recently repossessed (research 2015)" (Neilson and Castro 2016).

Coastal and Small-Scale Fishers: *Alive and Kicking*

Alison: It is difficult to include all these serious issues when teaching about fishing and the ocean, but it has always been the fishers who have told me about struggles and injustices toward other fishers far from the waters in which they themselves fish. From fishers, I have also learned stories of survival and thriving communities.

On a trip to Denmark by invitation from researchers working on social issues of small-scale fishers, I met Matilde, a young university-educated woman who had just bought a house based largely on the financial security she had from her job gutting fish in the coastal community of Thorupstrand. I was intrigued to know why her story was different from the ones I know in the Azores, other parts of Portugal, Canada and around the world, where systems of fish quotas which can be bought and sold for profit far from the sea were decimating and depopulating coastal fishing communities as the fishers, and especially their children, cannot afford to buy back their rights to fish and only large industrial fishing can survive in this system.

For the past 1000 years, the Danish fishing sector has been characterized by equal access to the resource, promoting share-organized fishing, where boat-owners and crew members share the earning from their catch equally. Thorupstrand, on the coast of Skagerrak, Denmark, is one of the last places on the North West coast of Jutland where fishers land their catch on the beach (Højrup 2011). By ensuring excellent quality of fish as well as rapid transportation from sea to market, they sell the fish at a premium price. Here, young people still engage in low impact fisheries which protect ocean ecosystems, maintain lived relationships between people and fish, and by valuing the principles of equity and cooperation, are managing to build hopeful futures where they can afford to own homes and provide comfortable conditions for raising families (Andresen and Højrup 2012).

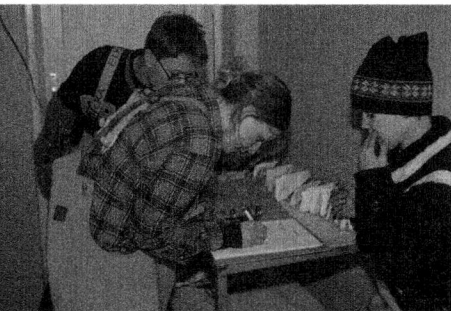

The industrious youngsters are called "the Gold of Thorupstrand" and they are able to gut 30 tons of plaice on a busy evening. They are working and fighting for their own future as sea people and the future of their coastal community. (Photo: Thomas Højrup 2008 (Højrup 2011))

The 2007 initiated system of Individual Transferable Quotas (ITQ) created disparities between the formerly equal share fishers: the share-organized fishers who did not own boats were disadvantaged as the quota was given only to the boat-owners ("to" the boat). The amount of quota was distributed on the basis of a three-year catch history which the crew members (fishers without boats) had taken part in creating, but whose part was not recognized. Additionally, the transferability of the quota (fishing rights) to increasingly fewer investors, made it almost economically impossible for young people to buy quota to be able to enter the fishing industry (Høst 2015). In Thorupstrand, during the introduction of the ITQ system, fishers collectively worked together to form a cooperative and together bought quota that they own in common. In 2014 small-scale fishers across Denmark created *Forening for skånsomt kystfiskeri*, an Association for Low-Impact (Sustainable) Coastal Fishing to help protect their interests against large trawlers taking over the entire fishing industry (personal communication M. Højrup, Nov 28, 2016).

Unfortunately, when I contacted Matilde for update for this chapter, I found out that the financial crisis, including the bankruptcy of the guild's bank with the resulting transfer of their loan, has been posing challenges to the survival of the quota cooperative. Financial instability continues as fish prices decreased, although the 2015 and 2016 increase in fish prices is giving renewed hopes for the future.

Rain Storm, Nazaré

When the sun goes down in Nazaré the women wrap woollen shawls around their shoulders and breasts, and tuck the ends into the waistbands of billowing skirts. Scarves are adjusted around the ears and tucked into the neckline of blouses. Wool socks are hauled up over large-veined legs.

When rain comes, heads are bowed into it, and shoulder blades buckle under it. Children are called to in high-pitched wails. Wooden shutters are banged to. Newsstands folded in. Dogs run sideways into doorways only to be kicked out again into the torrent.

Then, as abruptly as it began, it stops. Canaries sing and parrots chatter. Children run. Fish sellers yell. Men return to the damp sand and the repairing of long nets.

The sun goes down like a herald. The dogs overturn garbage cans. Children are called home. Sardines are placed on outdoor fires. Women sing. The sun sets. (Agnes Walsh 2009, used with permission of author)

Rita: I remembered seeing the seven skirted women of Nazaré selling fish when I went to the beach. According to GLOBEFISH, the unit in the Fisheries Department of the Food and Agriculture Organization of the United Nations (FAO), one in every two seafood workers is a woman, yet for the most part, they remain invisible, unrecognized and unacknowledged by the seafood industry in most countries. Although women participate in all segments of the seafood industry, including fishing, farming, trading and selling, monitoring and administration, decision-makers largely ignore them (Monfort 2015). When I first looked through the photos and the reports from your past work with fishing communities, I noticed the presence of women. Their importance in fisheries is recognized by communities around the world, and through networks such as AKTEA European Network of Women in Fisheries and Aquaculture which was created in 2006. This network of organizations from eleven different European countries promotes women's rights and women's visibility, specifically directed toward participation in decision-making processes in fisheries governance and the assignment of a legal status to all women working in family fishing businesses. They have had successes at the European Parliament gaining voice and recognition publicly at institutional levels, yet their work suffers from lack of funding.

Ilhas em Rede, an association of women fishers from all 9 Azorean islands was created in 2008 to promote women in fisheries including training, exchanges, participation in regional, national and international events and policy discussions. In their 2016 annual anniversary meeting, they highlighted their focus on three challenges: (1) safety at sea, (2) sustainability of fisheries, and (3) appreciation of women in fishing
Cristina Marques, São Miguel Island; Ângela Rodrigues, Pico Island; Fatima Garcia, Faial Island (Sempere 2008)

www.facebook.com/ilhas.emrede

Having just returned from a fisheries meeting, where we confronted again the stark reality of the lack of social science, my conviction for us to launch a network for social issues related to coastal fishing communities in the Azores is reinforced, especially when I look toward women's and other fishers associations to bring together small-scale fishers, university and community organizers to support research, development and policy creation. Other networks exist, such as *Too Big To Ignore (TBTI)*, a research network and knowledge mobilization partnership established to elevate the profile of small-scale fisheries, to argue against their marginalization in national and international policies, and to develop research and governance capacity to address global fisheries challenges. TBTI was initiated and is managed from Canada and has 15 partners and 62 researchers from 27 countries around the world (http://toobigtoignore.net/). *Low Impact Fishers of Europe (LIFE)* was launched in 2013 at the First Artisanal Fisher's Congress and enables fishers to develop and communicate collective positions and to influence policy, including the EU Common Fisheries Policy (O'Riordan 2016).

Implications for Education

Alison: With all the complexities of trying to sustain fish and fishing around the world, it is probably not surprising that I am critical of simple technical solutions such as farming fish especially based on the assumption that it is too late to protect wild fish. Concerns about aquaculture have been written by fisheries biologists (Rigby et al. 2016) as well as other environmental scientists (Young and Matthews 2010), but I am particularly concerned about how this becomes known to educators and promoted via formal and informal education, especially in the context of conflicting constructions of sea as nature (Braun and Castree 1998) and the validity of narratives that suggest that coastal communities of small-scale fishers are inevitably bound for extinction, and that aquaculture is the only way to provide current and future global food security. Similarly, unexamined assumptions, that privatization of the legal right to fish is the best or only way to protect sea life, are threats to the sustainability of wild fish and to the survival of communities of people who have the best experience and possibilities to live in tune with ocean ecosystems. During research to explore the ways that different people understood the sea, I was troubled by how fishers were perceived by educators who themselves worked or did leisure activities on the sea. Some even complained about the fishers not wanting to share the space with them, suggesting that fish were free for the taking, unlike farming which required work to create the food, and that fishers were irresponsible with both fish and money. The depth of our faith that we can and should control our world instead of learning to live within ecosystems which have functioned since the beginning of time is great.

Meeting in Rabo do Peixe, São Miguel Island between members of Ilhas em Rede and university researchers from multiple countries as part of Exploring the wealth of coastal fisheries: Listening to community voices, 21–24 October 2011 (Bulhão Pato et al. 2011). (Photo: Laurinda Sousa 2011)

Rita: The environmental classism reinforces a nature/culture divide which ends up enacting violence especially against children of coastal fishers since overtly and covertly the message taught is that people are always bad. Research in historical practices suggests that humans are clearly an integral part of ecosystems (Berkes 2004) having co-evolved with interdependent and affective connections between humans and non-human nature; "depicting 'humans' as the natural enemies of wilderness is thus irrespective of the millions who have been and continue to be victims of enclosures and 'improvements', a never-ending historical process that four decades of global neo-liberal politics and trade have reinvigorated" (Barca 2014). Fisheries education with goals to "train fisheries professionals with the technical knowledge and practical skills" and to "cultivate more environmentally knowledgeable citizens" (Crook and Zint 1998), tell us that the dominant explanation of problems of fishing are based on apolitical understandings, and the dominant solutions rely on the neutrality of knowledge and technologies, which are transmitted one-way from expert to ignorant fisher.

The sea was a treasure for me, because I collected lapas, caught lots of fish with a cane sitting on the rocks, caught parrot fish, caught triggerfish, caught young horse mackerel, caught horse mackerel, caught everything that was there. I was crazy. It was my best neighbour, my best friend and I exploited it, because I did very well in exploiting it. When I left my gate, a neighbour of mine would tell me: "You are not going alone to the coast, are you?" And I said: "Yes. But I never go alone." "Who are you going with?" "I'm here with Our Lord." When I go out the gate I say, "Our Lord come with me and I pay your alms." I have nothing thee, but I always come home loaded. I have enough for me and I have some to give away.[4] (Matilde do Coração Jesus, Santa António, Pico Island, Azores, 2009 interview)

AMPA, the Association of wives of fishers and shipowners of Terceira Island, with the collaboration of local fishers have embarked on a project of fishing tourism to create additional income as well as educate about small-scale fishing in the Azores Islands. Small groups can learn about artisanal fishing techniques via hands-on experience with the fishers, with the opportunity to engage in discussions about the day-to-day operations as well as the intricacies of policies and regulations before enjoying the taste of freshly caught fish in the restaurant of the local fishing village. (Photos: Alison Neilson 2010)

We invite educators to look toward and listen for the voices of the people who live in the oceans when exploring issues with students. The issues are complex and often hidden from sight when only considered from the narrow lens of natural sciences. Many of the "facts" told within the biology of the sea stories, are in fact, strongly contested, even by those people who also share the desire to maintain abundant fish life in the sea. We are reminded of the words of dian marino, *to be passionately aware that you could be completely wrong* (1997). In the midst of call for critical reflection, we hope you too can dream about a not-too-distant future where we could eat fish knowing that there are healthy populations in the ocean and where fishing communities could continue to exist in a sustainable manner with high levels of well-being.

---

[4] o mar foi uma riqueza para mim, porque eu, era lapas, apanhava muito peixe com uma cana sentada no calhau, apanhava vejas, apanhava peixes-porcos, apanhava carapau, apanhava charro, apanhava tudo o que havia. Era danada. Foi o meu melhor vizinho, o meu melhor amigo e eu fui judia dele, porque soube muito bem explorá-lo. Quando saía pelo portão fora havia uma vizinha minha que me dizia: Tu não vais sozinha para a costa? E eu disse: *Vou. Mas nunca vou sozinha.* Com quem é que tu vais? Vou aqui com o Nosso Senhor. Quando saio pelo portão fora digo, Nosso Senhor que vá comigo que eu pago as suas esmolas. Eu não tenho nada lá, mas venho sempre carregada para casa. Tenho para mim e tenho para dar (Matilde do Coração Jesus)

References

Abecasis, R. C., Longnecker, N., Schmidt, L., & Clifton, J. (2013). Marine conservation in remote small island settings: Factors influencing marine protected area establishment in the Azores. *Marine Policy, 40*, 1–9. https://doi.org/10.1016/j.marpol.2012.12.032.

Andresen, J., & Højrup, T. (2012). An alternative solution to the tragedy of enclosure. Experiences from fishermen's development of a common community quota in Denmark. In K. Schriewer & T. Højrup (Eds.), *European fisheries at a tipping-point* (pp. 303–366). Murcia: Cátedra Jean Monnet Universidade de Murcia.

Barca, S. (2014, October 18). Discussant for Wilderness Act forum. An environmental history online forum. http://environmentalhistory.net/wilderness-act-forum/

Béné, C., Arthur, R., Norbury, H., Allison, E. H., Beveridge, M., Bush, S., et al. (2016). Contribution of fisheries and aquaculture to food security and poverty reduction: Assessing the current evidence. *World Development, 79*, 177–196. https://doi.org/10.1016/j.worlddev.2015.11.007.

Berkes, F. (2004). Rethinking community-based conservation. *Conservation Biology, 18*, 621–630. https://doi.org/10.1111/j.1523-1739.2004.00077.x.

Braun, B., & Castree, N. (1998). *Remaking reality: Nature at the millennium*. London: Routledge.

Bulhão Pato, C., Neilson. A., & Sousa, L. (2011, October 21–24). *Exploring the wealth of coastal fisheries: Listening to community voices*. Unpublished final report. Angra do Heroísmo and Ponta Delgada, Portugal. http://conferencewealthofcoastalfisheriespt.blogspot.pt/

Carvalho, N., Edwards-Jones, G., & Isidro, E. J. (2011). Defining scale in fisheries: Small versus large-scale fishing operations in the Azores. *Fisheries Research, 109*, 360–369. https://doi.org/10.1016/j.fishres.2011.03.006.

Chubb, J., & Watermeyer, R. (2016). Artifice or integrity in the marketization of research impact? Investigating the moral economy of (pathways to) impact statements within research funding proposals in the UK and Australia. *Studies in Higher Education*, 1–13 (online). https://doi.org/10.1080/03075079.2016.1144182.

Cole, S. C. (1990). Cod, God, country and family. The Portuguese Newfoundland Cod fishery. *MAST, 3*, 1–29.

Converse, P. E. (1972). Change in the American electorate. In A. Campbell & P. E. Converse (Eds.), *The human meaning of social change* (pp. 263–337). New York: Sage.

Crook, A., & Zint, M. (1998). *Guide to fisheries education resources for grades K-12*. Maryland: American Fisheries Society.

Einarsson, N. (1993). Environmental arguments and the survival of small-scale fishing in Iceland. In G. Dahl (Ed.), *Green arguments and local subsistence* (pp. 117–128). Stockholm: Stockholm University.

Ertör, I., & Ortega, M. (2015). Political lessons from early warnings: Marine finfish aquaculture conflicts in Europe. *Marine Policy, 51*, 202–210. https://doi.org/10.1016/j.marpol.2014.07.018.

European Union and European Academies' Science Advisory Council. (2016). *Marine sustainability in an age of changing oceans and seas* (EASAC policy report 28). Luxembourg: Publication Office of the European Union.

Gough, N. (1993). Narrative inquiry and critical pragmatism: Liberating research in environmental education. In R. Mrazek (Ed.), *Alternative paradigms in environmental education research*. Troy: North American Association for Environmental Education. www.edu.uleth.ca/ciccte/naceer.pgs/pubpro.pgs/alternate/14.Gough.rev.htm. Accessed 18 Sept 2000.

Governo dos Açores. (2015, April). *Melhor pesca, mais rendimento medidas estratégicas para o setor pesca dos açores, 2015–2020* (Better fishing, more income strategic measures for the sector of the Azores). Available online https://www.azores.gov.pt/NR/rdonlyres/C3FD7DA5-9A78-415F-B541-856BACF411E6/0/PGRPlanoAcaoParaAumentarRendimentoPescadoresCRP.pdf

Hardin, G. (1968). The tragedy of the commons. *Science, 162*, 1243–1248.

Hind, E. J. (2015). A review of the past, the present, and the future of fishers' knowledge research : A challenge to established fisheries science. *ICES Journal of Marine Science, 72*, 341–358. https://doi.org/10.1093/icesjms/fsu169.

Højrup, T. (2003). State, culture and life-modes. In *The foundations of life-mode analysis*. Aldershot: Ashgate.

Højrup, T. (2011). *The need for common goods for coastal communities*. Gylling: Narayana Press.

Høst, J. (2010, October 18–22). *A neoliberal catch: Access rights and the clash of coastal lifemodes*. Presentation at the World Small-Scale Fisheries Congress, Bangkok, Thailand.

Høst, J. (2015). *Market-based fisheries management*. London: Centre for Maritime Research, MARE & Springer.

Jentoft, S., Chuenpagdee, R., Bundy, A., & Mahon, R. (2010). Pyramids and roses: Alternative images for the governance of fisheries systems. *Marine Policy, 34*, 1315–1321. https://doi.org/10.1016/j.marpol.2010.06.004.

King, T. (2003). *The truth about stories: A native narrative*. Toronto: House of Anansi Press.

King, T. J. (2005). Crisis of meanings: Divergent experiences and perceptions of the marine environment in Victoria, Australia. *The Australian Journal of Anthropology, 16*, 350–365. https://doi.org/10.1111/j.1835-9310.2005.tb00316.x.

Lam, M. E., & Pauly, D. (2010). Who is right to fish? Evolving a social contract for ethical fisheries. *Ecology and Society, 15*, 16. https://doi.org/10.5751/es-03321-150316.

marino, d. (1997). *Wild garden. Art, education and the culture of resistance*. Toronto: Between the Lines.

Merchant, C. (1997). Fish first! The changing ethics of ecosystem management. *Human Ecology Review, 4*, 25–30.

Monfort, M. C. (2015). Fishing out the invisible. *Samudra Report, 71*, 31–34.

Neilson, A. L., & Castro, I. (2016). Reflexive research and education for sustainable development with coastal fishing communities in the Azores islands: A theatre for questions. In P. Castro, U. M. Azeiteiro, P. Bacelar Nicolau, W. Leal Filho, & A. M. Azul (Eds.), *Biodiversity and education for sustainable development* (pp. 203–217). Dordrecht: Springer. https://doi.org/10.1007/978-3-319-32318-3_13.

Neilson, A. L., Bulhão Pato, C., & Sousa, L. (2012). A short reflection on research and fishing cultures performing knowledge together /Uma breve reflexão sobre o modo como investigadores e pescadores podem cooperar pelo conhecimento. *Revista Maria Scientia, 2012*, 73–82.

Neilson, A. L., Bulhão Pato, C., Gabriel, R., Arroz, A. M., Mendonça, E., & Picanço, A. (2016). In the Azores, looking for the regions of knowing. *Island Studies Journal, 11*(1), 35–56.

Neves-Graça, K. (2004). Revisiting the tragedy of the commons: Ecological dilemmas of whale watching in the Azores. *Human Organization, 63*, 289–300.

O'Riordan, B. (2016). New, but long overdue. *Samudra Report, 72*, 20–22.

Olsson, P., Folke, C., & Berkes, F. (2004). Adaptive comanagement for building resilience in social-ecological systems. *Environmental Management, 34*, 75–90.

Pierce, G., Pita, C., Santos, B., & Seixas, S. (2012). Sustainability of fisheries. In F. J. Goncalves, R. Pereira, W. Leal Filho, & U. Miranda Azeiteiro (Eds.), *Contributions to the UN decade of education for sustainable development* (pp. 325–367). Frankfort: Peter Lang.

Rigby, B., Davis, R., Bavington, D., & Baird, C. (2016). Industrial aquaculture and the politics of resignation. *Marine Policy* (first online). https://doi.org/10.1016/j.marpol.2016.10.016.

Rittel, H. W. J., & Webber, M. M. (1973). Dilemmas in a general theory of planning. *Policy Sciences, 4*, 155–169.

Rocklin, D. (2016), Who's who in small-scale fisheries. In R. Chuenpagdee & D. Rocklin (Eds.), *Small-scale fisheries of the world* (Vol. I, 8 pages). TBTI Publication Series: St John's.

Sempere, M. J. (Ed.). (2008). *Estamos cá. Existimos. As mulheres na pesca nos Açores*. Ponta Delgada: UMAR Açores.

Shackeroff, J. M., Hazen, E. L., & Crowder, L. B. (2009). The oceans as peopled seascapes. In K. McLeod & H. Leslie (Eds.), *Ecosystem-based management for the oceans* (pp. 33–54). Washington, DC: Island Press.

Symes, D., & Hoefnagel, E. (2010). Fisheries policy, research and the social sciences in Europe: Challenges for the 21st century. *Marine Policy, 34*, 268–275. https://doi.org/10.1016/j.marpol.2009.07.006.

Symes, D., Phillipson, J., & Salmi, P. (2015). Europe's coastal fisheries: Instability and the impacts of fisheries policy. *Sociologia Ruralis, 55,* 245–257. https://doi.org/10.1111/soru.12096.

Tacon, A. G. J., & Metian, M. (2008). Global overview on the use of fish meal and fish oil in industrially compounded aqua-feeds. Trends and future prospects. *Aquaculture, 285,* 146–158. https://doi.org/10.1016/j.aquaculture.2008.08.015.

Vlachopoulou, E. I. (2016). *Fishing for human lives. Impacts of the refugee crisis on the fishers of Eastern Greece.* Poster presentation for *TBTI Symposium on European Small-Scale Fisheries and Global Linkages,* Universidad de La Laguna and Oceanographic Centre of Canary Islands, Santa Cruz de Tenerife, 29 June to 1 July, 2016.

Walsh, A. (2009). *Para o Santa Maria Manuela.* Translated into Portuguese by Paulo da Costa, Letterpress-printed and hand-sewn limited edition. Running the Goat Press and City of St. John's, Newfoundland and Labrador.

Young, N., & Matthews, R. (2010). *The aquaculture controversy in Canada: Activism, policy, and contested science.* Vancouver/Toronto: UBC Press.

Alison Laurie Neilson is a transdisciplinary social scientist working on environmental justice in fishing communities of the Azores Islands. She uses narrative and arts-informed research on the way sustainability is understood and manifested in education and policy, how people learn to be part of the governance system and how the processes of education construct the issues and structures.

Rita São Marcos is a sociologist who has participated in research and community outreach projects focused on the governance of environmental issues in the Azores. As a PhD candidate, supervised by Alison Neilson, she is currently exploring the tensions between participatory democracy, expert knowledge and small-scale fishers' participation in the EU Common Fisheries Policy.

Chapter 8
Gleaning White-Tailed Deer

James Farmer

During my senior year at Indiana University, I lived out in the countryside by Lake Monroe – the large reservoir that provides water to Bloomington, Indiana – in a small log cabin with poplar logs hewed in the nineteenth century. I had moved to the cabin to re-connect with wildness, and have since lived all but 2 years out of 21 in rural, wooded landscapes. Driving home from town on an October Friday night I came upon the flashing lights of a cop car on HWY 446. There was only one car at the scene of an accident, and as I slowly passed by I realized that the vehicle had struck and killed a white-tailed buck. I had a taste for venison, so in a moment of spontaneity I stopped to see what was going to happen with the animal. What a gift! I pulled back onto the highway with a whole deer in my vehicle and went home to figure out how to butcher it. The process woke me up, providing an authentic connection to the land.

Natural History Lesson

I was driving my T-100 on Hoover Road one morning when I hit a yearling on the short stretch of road between Morgan's and Wilkerson's Sawmills. The deer was paralyzed. The only tool I had was an end splash of a cultured-marble vanity top, the kind of fake marble used in bathrooms in the 1980s. The piece was about 4 inches by 20 inches. I thought about it for a second, grabbed the chunk of material and used it as blunt club to strike the deer at the base of the skull where it attaches to the spine. Two forceful thumps were enough to kill the poor animal.

J. Farmer (✉)
School of Public & Environmental Affairs, Indiana University, Bloomington, IN, USA
e-mail: jafarmer@indiana.edu

© Springer Nature Switzerland AG 2020
J. B. Pontius et al. (eds.), *Place-based Learning for the Plate*, Environmental Discourses in Science Education 6,
https://doi.org/10.1007/978-3-030-42814-3_8

I proceeded to field dress the deer and then called my principal to ask if I could bring the animal to school for my at-risk students. It was an unexpected conversation, of course, but he agreed that it would be a good learning experience. The lesson plans on grammar and early American history were scrapped for the day and instead we set out to butcher a deer. While I was apprehensive to turn my students loose with kitchen knives, I did hand out small pieces of obsidian. In hindsight, we may have been safer with the kitchen knives. The lesson was simple. Remove the muscle groups slowly, one at a time to carefully study how the muscles work to move the bones in conjunction with tendons and ligaments.

This was without question the best day I ever had as a public school teacher. The experience completely captivated the students from start to finish. Students learned how a body works through direct experience. They watched with patience and worked with care. There were no mishaps or accidents, just enthusiasm and curiosity and wonder, as most had never been close to a deer, let alone butchered. I am sure the students remember this experience, which is more than I can say about other lessons I taught.

Spontaneous Steak

I remember hearing a story that President Jimmy Carter had mused that he, especially in challenging situations, "always felt comfortable knowing he was likely the only person in the room that ever had eaten opossum." I am unsure whether or not he truly uttered these words or if these words were more metaphoric than literal – but it seems to fit Carter's character. Regardless, I love the idea – it resonates with me. I often find similar comfort within the halls of academia knowing I am likely one of the few professors that gleans white-tailed deer from the rural roads around home.

Southern Indiana, where I have lived all but 3 of the last 23 years, has an incredible overpopulation of deer. They seem greatest in number when you hit the exurban and urban areas. Nevertheless, they are plentiful in the countryside. Personally, I have hit 4 deer in driving the windy roads that traverse the ridge tops and valleys in and around Brown County and Bloomington, Indiana. Not all were killed or even hurt. Where I live, in an area of densely-forested, hilly terrain, deer are often killed by vehicles. In fact I saw a deer on the roadside the day I started writing this chapter. I considered stopping for a closer look, but I had not been through the area for several days, so I had no way of being sure the meat was safe.

The most current stats on deer-elk-moose-vehicle collisions, the most common North American ungulates, estimate over 1.5 million accidents occurred between the two (State Farm Insurance 2016). The majority of these are with deer. Such accidents are most likely to occur if you are driving in West Virginia, Montana, Pennsylvania, Iowa, or South Dakota, with one's odds of hitting a large ungulate at 1 in 41, 58, 67, 68, and 70, respectively. In my home state of Indiana, the likelihood is 1 in 136 drivers will hit a deer annually (State Farm Insurance 2016). I suppose that is why I have hit four over the past 24 years. The bonus antlerless licenses available from the Indiana Department of Natural Resources are some indication of the deer population in Brown and Monroe Counties. You can take up to 8 extra antlerless deer in Monroe and 4 in Brown. Most deer-auto accidents occur between October through December, during mating season and around dawn or dusk when deer are most active. This makes traversing roads surrounded by woods or tall corn rows more likely a place for a collision to occur.

Deer have not always been so abundant on the midwestern landscape. Populations in the early 1800s were quite low due to over-harvesting by indigenous people. This trend continued into the late nineteenth century as market hunting continued to thrive, particularly as game could be shipped by rail. It was not until the Lacey Act (of 1900) that policy was placed on the illegally harvesting and distribution of wild game. However, white-tailed deer were extirpated across much of the landscape by this time. For example, by 1893, white-tailed deer were completely absent from the Indiana landscape (Indiana Department of Natural Resources 2017). With their re-introduction between 1934–1942, and the growth of edge habitat, they have rebounded and are estimated to be around 950,000 across the state (North American Whitetail 2014). This increase in number, decrease in hunters, elimination of predators, one-buck hunting rule, and development occurring at the rural-urban fringe have made a recipe for road-kill to be harvested.

I have been more of a gatherer of game than a hunter of game, not even really a gatherer of fruits, nuts, and other wild edibles. In most common occurrences when gatherer and hunter are phrased together, the word hunter generally comes before gatherer. Additionally, 'hunting' seems to imply game animals such as – in my region – squirrels, rabbits, turkeys, grouse, and deer, whereas 'gathering' relates more to fruits, nuts, mushrooms, and other wild edibles. 'Gleaning' is most often used to describe the collection of otherwise not-harvested, feral foods such as apples growing in abandoned orchards. How should I then classify my collecting of deer from the roadsides? I am not a part of a community of people who gather/glean/hunt for deer on the sides of the road, although I know there are others who do this.

Until late, the vast majority of road kill harvesting occurred in rural areas among people who are likely to have the skill, knowledge, and ability to make use of the animals, though the activity has increased in suburban and urban areas. A key reason for engaging in the activity is for food security purposes. Experts are quick to point out that rural individuals are more likely to suffer from food insecurity than individuals living in urban areas. The reasons are many, however, proximity to available food supplies coupled with transportation barriers are high on the list. Understanding the role of road-kill deer in food security, and its placement in rural culture, is easily traced by weaving it through the definition of food security.

People have diverse reasons for wanting to eat gleaned venison. A friend of mine considers herself part of a "vegans for venison" movement. Of course many vegans would be repulsed by the notion. However, those who are vegan for the sake of sustainability might consider venison as an alternative to farm-raised meat granting that wild venison for all intents is carbon neutral. With the continued overpopulation of white-tailed deer, their consumption may be understood as a sustainable alternative to procuring beef, lamb, goat, pork, or poultry from a grocer or even a local farmers' market.

Gleaning white-tailed deer can also be part of food security. Food security is defined by individuals having access to, the ability to acquire, and the knowledge to prepare foods that are healthful and allow them to live a healthy, active lifestyle. While not all people who glean from the roadsides are food insecure, the vehicular transportation routes do supply food insecure people with vital protein as do "hunters for the hungry programs" and other similar movements. Because deer have long been on the rural landscape, access to them is well established.

A common misconception is that the harvesting road-kill animals for human consumption is legal in all 50 states. This is far from the case. Based on a review of recent articles and conversations on the topic, only about 20 states legally allow the collection of road-kill animals for human usage. Indiana, luckily, is a state that has long-allowed the practice. The prominence of white tails in suburban and urban settings is increasing in frequency as discussed earlier in this chapter. Thus, deer-car collisions in these settings are becoming more common. To aid in this are road kill phone banks, a list kept by sheriff departments and animal control units who are able to communicate with interested parties to utilize dispatched wildlife. Many counties keep a call log of individuals seeking road kill if the driver of the colliding vehicle is disinterested in claiming the carcass.

Gleaning Skills

The ability to acquire road kill is not merely an issue of economics, but it also includes the capacity to butcher the animal before one even considers preparing it as food. This type of knowledge was *generally* communicated via direct experience and passed through generations as young boys were invited to join the hunting parties of their older, male familial counterparts. Hunting in the United States and many western cultures has long been a male dominated, and heavily stereotyped activity. In Indiana, the opening weekend (usually the 2nd weekend of November) of deer season and Thanksgiving Day are classic dates for such group hunts to occur. Opening weekend of deer season in southern Indiana is littered with makeshift camps as hunters flock to public lands (state and national forests), leased lots, or a friend/relative's property for a weekend with nature.

While the Thanksgiving Day hunt tradition was not one my family engaged in, I did partake in the ritual with a close-friend's family and that is where I learned some of my gleaning skills. These are key occasions where the knowledge and skills of field dressing a deer, inspecting the carcass for disease (blue tongue, chronic wasting disease, etc.), possibly butchering the carcass, and telling a good story are shared. Of the four previously listed skills, butchering is the one that is being lost as more individuals hand the job to a local processor who usually performs that task for a $50–$100 fee.

I learned to field dress and butcher deer from Bob and Shorty Smith, uncles to my friend Mark. Mark and I showed up at their farm with two whole animals one day. Neither of us had a clue what to do. We first removed the entrails by making an incision from groin area up to the base of the sternum- making sure to not puncture the bladder cavity, intestines, or rumen. We then removed the entrails and other organs, all the while trying to keep the urine and fecal matter contained in their organs. At the time I was not into organ meat. Finally, we washed out the chest/torso cavity, removed the tenderloins that run along the spinal column on the inside of the ribcage for the night's dinner, and hung the animals in the barn. Mark and I went back three days later for a lesson on butchering.

We started in the old milk house of what used to be a dairy farm and began by skinning out the deer. This was followed by our use of a small, hand-held propane torch to singe the random hairs that fell of the hide and onto the meat. Our approach was simple, and fairly crude now looking back. We dissected the carcasses trying to remove muscles in their entirety, washed them, and wrapped them up in white butcher paper. The scraps of meat that resulted from the process were collected and taken to a processor where they would make summer sausage, another skillset unto itself.

To a lesser extent than making summer sausage, to enjoy eating white-tailed deer, one must develop the ability to prepare and serve the meat. Like other types of meat, the preparation and the health and age of the individual animal is critical to its palatability. The whole process takes so much work that I, at least, feel it is important for the food to turn out as well as it can.

An old deer, like an older rooster, will make for tough, stringy meat. If we end up with an older deer, we generally grind this into burger or sausage. However, younger white-tailed deer can be delicious in the form of steaks, kabobs, roasts, or tender stews. I generally marinate larger cuts of meat with olive oil and garlic as it gets consumed as steak or roast. The larger hunks, once cut up, are also excellent for use in fajitas, burritos, or chili. I do not recall ever having bad tasting venison, just tough meat that is difficult to chew.

Gleaning for Leisure

Apart from harvesting of road-killed deer from sustainability or food security purposes, others may engage as a leisure pursuit. Leisure has a plethora of meanings. Merriam Webster defines leisure as, "freedom provided by the cessation of activities; *especially*: time free from work or duties," (Merriam-Webster 2017). Most seem to equate it to said activities, or even worse yet, to idle time or some*thing* arbitrarily pursued when no other obligations need to be met. Leisure scholars have a differing view. While several definitions are nuanced to specific leisure theorists, I more or less subscribe to Geof Godbey's description of leisure as "living in relative freedom from the external compulsive forces of one's culture and physical environment so as to be able to act from internally compelling love in ways which are personally pleasing, intuitively worthwhile, and provide a basis for faith," (Godbey 1985). This is to say that leisure, true leisure, is the pursuit of one's passions not because of what friends, government, or mother culture tells you to do, but for intrinsic reasons. Pursuit of leisure provides worth and value in and of itself: gratification is received in leisurely experience in a deeper way.

While leisure has existed to greater or lesser extents throughout much of human natural history, the opportunity for people to engage in leisure (speaking mostly to westerners) is primarily a manifestation of the twentieth and twenty-first centuries. Not until time and resources were of surplus did individuals have the capacity to pursue leisure. However, during this same period (albeit not nearly as much during periods of say the Great Depression or World War II) American culture broadly shifted from a community orientation and reliance on neighbors, to an individualistic culture focused on property rights and heading toward neoliberalism. The lack of want and availability of surplus allows leisure pursuits.

During this time period (1880s-present), people simultaneously migrated in hoards to urban areas. By 1950 over half the U.S. population lived in cities, and by 2010 the number had grown to roughly 80%. Individuals adopted the city life, spending more time in front of televisions and smart-phones, days mowing fescue lawns, driving children around in mini-vans to baseball tournaments and soccer matches, and disconnecting with nature. This disconnect and desire (and need) to connect with nature has become the focus of much contemporary scholarship. Starting with Edward O. Wilson's *Biophilia* proposition (1980), to Louise Chawla's (1998) scholarly work on significant life experiences with nature Joy James and Rob Bixler's conceptualization of environmental socialization (2016), and popularized by Richard Louv's *Last Child in the Woods* (2005), American scholars and journalists are penning about (1) the seriousness of our disconnect with nature and (2) our innate desire to connect with nature.

Nearly 40 years ago Wilson, an ecologist, described the human innate desire to have a relationship with, to know, and to grow in and with nature. He termed the notion "biophilia," and describes this emotional connection between humans and nature as an imperative that will be replaced by whatever environment is familiar. Thus, those growing up in and around wildness and natural landscapes will substitute suburban lawns and concrete horizons as the archetype impression. Chawla, alternatively, has built a line of scholarship that highlights the critical role that a *significant life experience* in, about, and/or with nature has on an individual and their attitude and behavior directed towards nature. Spending free and unstructured time in nature, going to a nature-based summer camp, witnessing a neighborhood developing in a used-to-be wooded place, or living through a massive oil spill (think BP) align with those significant life experiences she has found to best connect people with a life-long commitment to nature. Where Wilson and Chawla leave off, James and Bixler begin. They describe *environmental socialization* as the means by which individuals develop a strong, healthy, and supportive relationship for and with the natural environments and nature. Essentially less time = less relationship = less care. While this may not be true for all, their research finds it is true for many.

I propose gleaning road-kill white tails to be both a leisure experience that facilitates a connection with the land and wildness that is unfortunately missing from modern life. Gleaning white tails is an act of self-reliance, an act of connection with the world outside of our control, an act behaving "in relative freedom from the external compulsive forces of one's culture." I don't think I take too much liberty in stating that the majority of folks would turn their nose up to a road-killed supper. This flies in the face of common social norms yet at the same time, intrigues audiences worldwide. A British reality show featuring retired entomologist Arthur Boyt solely focuses on the harvesting of road kill, while I have passed artists along state highways brushing a recent road kill onto a fresh canvas.

A few winters ago I gleaned a young button-buck just a hundred yards from my driveway. My wife had just arrived home from teaching a night class on the Indiana University Bloomington campus, and told me that a teenage driver had just hit a deer in front of the house. I grabbed a .22 rifle, headed out the door, and found a couple of drivers who were trying to figure out what to do. The paralyzed deer was trying to drag itself off the road by its front legs.

The middle-aged man, driving a big red truck, was holding a handgun. As I walked up he said that he just couldn't bring himself to kill the deer. I killed the deer and went back to the house, drove my hatchback down the driveway, drug the deer up on top of the hood, and drove him home. My wife, Sara, came out of the house as I was field dressing the deer.

"How can you just stick your hands into that animal with no gloves on?" she asked. I had to think about it. "I guess I've just accepted how this process works and how it feels."

My three children watched and I periodically showed them the deer's organs, like the heart and lungs.

I gave a lot of thought to the man in the red truck. Why couldn't he shoot the deer? I thought about his concern for other living creatures, and how we might both share such a concern but behave in entirely different ways. But I also wondered what his story was. Why did he have a gun in his truck and what had he used it for? Did he grow up out here in the eastern part of the county, was he merely driving by, or was he a transplant seeking out the rural life but having grown up in the urban? What does it mean that I unflinchingly collected the animal as dinner?

By shooting the deer, I felt, oddly, like I was not the deer's killer. The car was the killer – the animal just hadn't died yet. It would have and soon, either by hypothermia from laying on the January ground, from a pack of coyotes during the night (when it was still alive), or possibly by another passing car if he had not made it entirely off the road. The deer was soon to take its last breath. I just finished what was already in motion. My empathy is what spurred me to end the deer's suffering and my instinct to use the animal's meat.

Harvesting road-kill deer also provides a leisure experience for me by allowing the opportunity to engage with the land around me. The act also provides a chance to utilize food that may otherwise go wasted, at least by humans, and the opportunity to eat meat that is virtually carbon-neutral. Gleaning along the roads provides a connection to spontaneity that most of life no longer affords with harried schedules of modern living. I am here engaging in the conversion of a deer to food, considering the life lost, the lives supported, and how humans and nature are one.

References

Bixler, R. D., & James, J. J. (2016). Where the sidewalk ends: Pathways to nature-dependent leisure activities. In D. Kleiber & F. McGuire (Eds.), *Leisure and human development* (pp. 107–131). Urbana: Sagamore Publishing.

Chawla, L. (1998). Significant life experiences revisited: A review of research on sources of environmental sensitivity. *Environmental Education Research, 4*(4), 369–382. https://doi.org/10.1080/1350462980040402.

Godbey, G. (1985). *Leisure in your life: An exploration* (2nd ed.). State College: Venture.

Indiana Department of Natural Resources. (2017). *Indiana deer biology.* Retrieved from https://secure.in.gov/dnr/fishwild/3359.htm

Louv, R. (2005). *Last child in the woods: Saving our children from nature-deficit disorder.* Chapel Hill: Algonquin Books.

Merriam-Webster. (2017). *Leisure.* Retrieved from https://www.merriam-webster.com/dictionary/leisure

North American Whitetail. (2014). *20 best whitetail states for 2014.* Retrieved from: http://www.northamericanwhitetail.com/trophy-bucks/best-whitetail-states-2014/

State Farm Insurance. (2016). *LOOK OUT! Deer damage can be costly!* Retrieved from: https://newsroom.statefarm.com/state-farm-releases-2016-deer-collision-data/#Teo6DjYlQr1CZJds.97

Wilson, E. (1980). *Biophilia.* Cambridge, MA: Harvard University Press.

James Farmer is an associate professor in the School of Public and Environmental Affairs at Indiana University-Bloomington. His research is focused on local and regional food system change and development. Specifically, he studies human behavior and decision-making related to farm diversification, participation in local and regional food systems, and food sovereignty. Farmer teaches courses on sustainable food systems and community resilience. He also directs the IU Campus Farm and convenes the Sustainable Food Systems Science initiative at Indiana University. James lives on a small farm in Unionville, IN with his partner and three children.

Chapter 9
My Father's Place in the Mountains: An Education Elsewhere

Amy Price Azano

I remember considering Kim Donehower's title question: *Why not school?* (Donehower 2013) in thinking how rural literacies intersect with the types of sustainable educational practices occurring outside of the school day—that is the rural literacies that both promote sustainability for rural places while also resisting the invisible pull of outmigration for rural learners. Donehower came to that question from interviewing "hyperliterates—people who read and/or write extensively as adults, when those activities are not required by a job or pursuit of a formal educational degree" (p. 35). What types of "hyperliterates" do rural communities foster? In asking that question for myself, I thought of my own hyperliterate practices—and my father's (the subject of this inquiry)—as a way of framing how literacies are actualized in rural places. I have never seen my father, a first generation high school graduate, read a book for pleasure; however, he embodies community-based hyper-rural literacies as they relate to hunting and to his group membership in the community where I was raised and where he still resides.

Using autoethnography and narrative inquiry, I sought to understand how rural literacies are actualized in rural communities and, as in my father's case, the ways one attaches meaning to the places in which those rural literacies are learned. In doing so, the case itself becomes rural literacy praxis, an opportunity to embody the practice I seek to investigate. In the ways that Eppley (2013) argued that rural literacies account for an inquiry into a county Fair, so too I hope will the story of my father's relationship with hunting.

A. P. Azano (✉)
School of Education, Virginia Tech, Blacksburg, VA, USA
e-mail: azano@vt.edu

© Springer Nature Switzerland AG 2020
J. B. Pontius et al. (eds.), *Place-based Learning for the Plate*, Environmental Discourses in Science Education 6,
https://doi.org/10.1007/978-3-030-42814-3_9

The Place of Memory: *Context for Inquiry*

Making the turn onto Cumbia Avenue, a gravel road belonging only to us and old Mrs. Vines, I could see my Dad's '72 Chevy pickup backed into the driveway. I knew what to expect—the deer tied up with rope and hanging from the attic door on the carport. It would be cut down the center and gutted, with a bucket and newspapers underneath to catch the blood. Sometimes there'd be another deer waiting on the back of the truck—belonging to my Uncle Walt, who would soon come over with beer to help. My Mom, working nearby at the picnic table, would be rationing out what she needed for meatloaf or pepper steak, and wrapping the rest for the freezer. I had never seen *Bambi*, but even if I had, I'm not sure the sight of the deer and the rifles, the camouflage coats and hats strewn about, or the empty coolers and beer cans, would have bothered me. The only thing I cared about was that Daddy was home from the mountains. He was alive and well, and we'd soon be laughing at one of his exaggerated stories over dinner (Fig. 9.1).

My Dad and I have always been close, but I'm much less of a "Daddy's little girl" and more like my father's only son. While my older sister was made of cotton candy and pom poms, I was from cotton—the wild child, the strong willed, the rugrat, the tomboy—and my father would have it no other way! We camped and fished, and he took me on adventures up to the gravel pit to shoot tin cans. (I was a

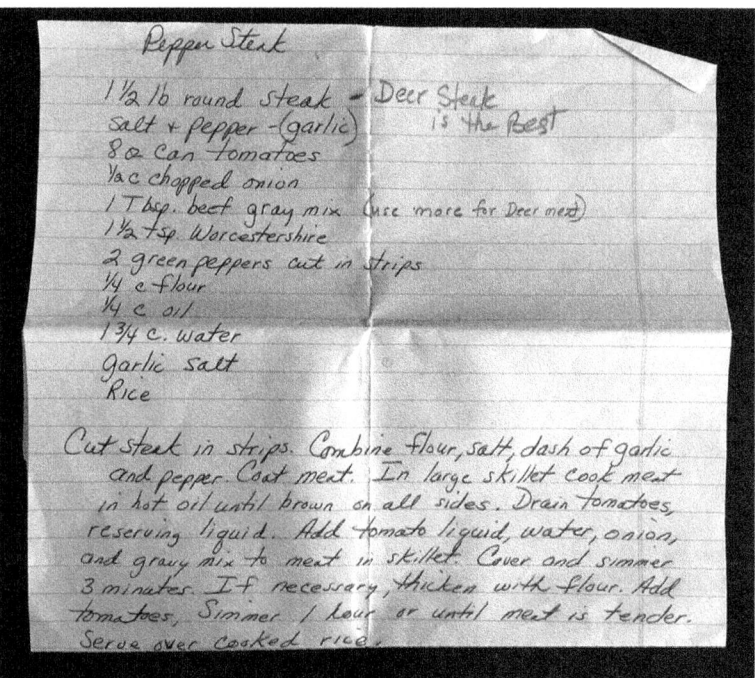

Fig. 9.1 Mom's pepper steak recipe

very good shot for a 10 year old!) He'd wonder aloud, "Do you think my Jeep can drive up the creek bed?" We had our answer a few hours later after we stalled out and walked in waist high water to the bank. He taught me how to grill steaks, to fix most anything, to drive a riding lawnmower, then a motorbike, and finally a manual transmission by the time I was 13. My college graduation gift was a beautiful metal toolbox complete with pliers, wrenches, screwdrivers, and the like. A few years later he bought me my first drill. When I'm making a repair or tackling some small carpentry job around the house, my son will ask: "How'd you learn to fix everything?" And I always answer, "My Dad taught me."

It's true that my father has taught me many lessons. As my first teacher, his greatest lesson was telling me that I could never do anything so bad that he hadn't already done. That opened a door I've had to walk through on more than one occasion—one where I could be honest about my mistakes. I often got in trouble, but (impossibly) I did so without being judged. My father is a flawed and deeply sincere man aware of his own transgressions in this world, using much of his adult life to make peace with his own past, his time in Vietnam, and his relationships—but where he might have regret or mis-stepped in other areas of his life, he never did so as a father. My Dad never missed a basketball or volleyball game, was always there with a camera when I donned the societal expectations of homecoming and prom gowns, never turned down an opportunity to talk or listen, tuck me in, and sat at the breakfast table with me each morning before school. If he raised his voice in frustration, he apologized. And after threatening many times to give me an M&M (a "mashed mouth") for my sass, he succumbed only once—that night I fell asleep with him crying at my bedside.

Perhaps it was the fact that he was always there that made his disappearing act at the start of hunting season so confusing. There were other disappearances, but those were before my memories take shape and come to me now like scenes from a forgotten movie—his coming home late after a night at the Moose Lodge, my parents arguing in the kitchen, whispered yelling so as not to wake me and my sister. I remember peaking at them on one of those nights. He had dropped a carton of milk and was cleaning up the spill on the green linoleum. The light of the refrigerator casting a spotlight on the felt red poppy pinned to his shirt. Memorial or Veteran's Day—holidays that still haunt my father.

Another one of those nights plays in its entirety. It begins the same with my Dad coming home late. My sister and I huddled under the covers of her bed, while our mother's damning criticism mixed with my Dad's drunken footfalls heavy in the hallway of our small, brick rancher. He stumbled and fell into their bedroom closet, knocking the louvered, bi-fold closet doors off their tracks. This followed by our mother barking for us to get in the car. In our pajamas, my sister and I both crawled into the front seat of her orange, "bittersweet mist," 1971 Grand Sport Buick. I clutched my ragged and worn "Tigey," while my Mom mumbled, "Come on, Betsy," as she tried the ignition. We were too sleepy or confused to cry, but that didn't stop my Mom from doing so as my Dad came from the kitchen into the carport, standing where those deer would hang, and begged, *Please don't go. I'm sorry. I'll never go back.* Betsy had just turned over, and we all sat idle. Without turning off the car, my

Mom got out, and in the glow of the headlights, I thought they looked like movie stars as they hugged and cried. My Mom came back and turned off the ignition. My sister and I went back to bed, and my Dad never went back to the Moose Lodge. In fact for the remainder of my childhood the only time I can recall his absence is during hunting season.

I have observed my father with keen interest all my life—the way he commands a room, his ability to tell a joke or a tall tale and have an entire group hanging on his every word, his larger than life persona that is at once humble and confident. He is respected in his small, rural community, and has a reputation of being a principled man. It is known that nothing comes before his family—except maybe the first week of black powder.

Storytelling and Telling: *Methods for Inquiry*

In this chapter, I try to make sense of my father's relationship with hunting and the culture of hunting as a window into the opportunities for place-based learning. I rely on a critical pedagogy of place (Gruenewald 2003) and Paulo Freire's (1970/2009) *conscientização*—the idea of being critically awake to oppressive forces (in this case education)—as a theoretical frame to make meaning of my Dad's experiences, and mine as his daughter. While I set out to "tell" my Dad's story, I realized upon writing and revising that I too had much to learn as a daughter of the subject of this chapter. Like many qualitative or ethnographic inquiries, the transactions between me and my father, me and my writing, and so forth constructed the meanings I share. Upon reading the chapter, the editors of this book independently asked me about my own identity. David asked, "What meaning can you make of your identity as a (relatively) urbanized woman (i.e., class and gender issues) with rural hunting roots?" My initial thought was to scoff at the word "urbanized." I still live in Appalachia. I teach at a land-grant institution in a rural region of the state; yet, for as much as I want to *identify* as rural, I am keenly aware that my life experiences are vastly different than the rural ones of my parents—even though we come from the same *place*.

Like my Dad, I grew up in the country with a family who will jokingly (and, admittedly, proudly) call themselves "rednecks." Yet, I have dedicated my career to challenging stereotypes and assumptions associated with that provincial title. I find no pride in it. Each semester in one of the graduate classes I teach at Virginia Tech, I model a "privilege walk" with my students. As I participate in the questioning with the preservice teachers, I often experience dissonance between how I would have responded to questions about privilege as a teenager and how my own teenagers would respond to these same questions now. My Dad had several different jobs when I was younger—from driving a bread truck (his CB handle was "Dough Boy"—and mine was "Wood Woman" because I could carry as much wood down to the basement stove in winter as the men in my family!) to selling cars or insurance to working as a supervisor in a local manufacturing plant. There were lean years when we struggled to buy the basics and fat years when we traveled to Disney World or built an addition on to our house. In the end, it was a lean year that prompted the selling of my childhood home after which my parents moved into my maternal grandparents' house where they still reside. While we wanted for little, privilege was variable in my life and it wasn't until I left for college that I realized "middle class" in the country meant something very different for children of suburbanized or college-educated parents. And, for my children—their privilege (and my own) is undeniable. With classroom teachers and police officers in my extended family, having a "good" job has always been a better measure of success than the size of a paycheck; yet, still—as I write this, currently a tenured professor at a public university, my salary is more than my parents have ever made combined.

In an attempt to tell and *read* my Dad's story critically, I considered how a critical pedagogy of place can serve to "evaluate the appropriateness of our relationships to each other, and to our *socio-ecological* places" (Gruenewald 2003, p. 7). As such, I draw attention to these relationships, the missed opportunities in my father's education, and also how my Dad's constructions of place serve as a place-based *text* in my own life. While place-based education is largely associated with outdoor or environmental education and my Dad's experiences as a hunter is a fitting case study for that more narrow definition, a *critical* reading not only provides insight into my father's relationship to and with nature but also how that relationship influenced my own identity—both serving as opportunities to explore "the ways that power works through places" (p. 7). In other words, *place* is more that an empty (geographic) container that holds our histories and experiences. It *becomes* a place by the meanings we attach to it—an idea I circle back to at the end of this chapter.

For the purpose of a methodology, I asked my Dad to start writing in a journal about his relationship with hunting. I rarely saw my Dad engage in more traditionally academic tasks, like writing, though—when I did—I was always amazed. I believe strongly that had my father grown up in a different place or with more financial and social capital, he would have gone to college and perhaps found a career as a writer. As a deacon at our local church, he occasionally substituted for our pastor, and his sermons (the ones he would labor over throughout the week) were always my favorite. As part of a high school project, I asked my Dad if he would write down and share with me his experiences in Vietnam. I still remember his description of hearing screams at night and not being able to discern if they were the screams of humans or monkeys. I offer this just to say that I knew my father would be able to succeed at the task of writing about hunting.

After my initial request, he shared four, single-spaced notebook pages with me. His initial pages described "what" he loved about hunting. I followed up and asked him to think about "why" he loves it so much. I asked, "What did it do for you as a young man? What does it still give you? Confidence? Peace and quiet? Your own space? Connection to nature?" These prompts led to a few additional pages of writing, with several intervening conversations with my Dad about hunting and fishing. The quotes throughout this chapter are from both his writing and the interviews. The italicized words in this chapter represent my memories of past conversations or experiences. I shared multiple drafts of this chapter with both my parents as a form of member checking to be sure I had represented the memories as accurately as possible.

Stony Man: *A Metaphor in Being*

My Dad is something of a good ol' boy. He grew up poor and restless on the outskirts of Luray, Virginia, in an area known as Stony Man—named for the mountain, the second highest peak in the Shenandoah National Park and part of the Appalachian Trail. Aptly named for the rocky outcropping at its summit, it's also a fitting description for the men who grew up in these parts, most notably my grandfather, nicknamed "Old Stone Face" for his stern and stoic nature. As a World War II veteran, my grandfather believed that hard work made you into a man. The only son of three children, my father worked tirelessly at home and was responsible for taking out the slop jar, feeding the chickens, digging potatoes, mowing the yard, butchering hogs, working the garden, and shucking corn. He learned to drive at nine years old and would take his Granddaddy in their old Chevrolet pickup to the feed mill to have corn ground for hog feed. For extra money, he worked for neighbors doing much of the same, in addition to feeding goats and cattle, and gathering hay to load in the barn. Also, at a very young age, he learned to hunt.

His game of choice has always been deer. He is at once infatuated by their peaceful nature and his desire to hunt them—a reconciliation that eludes many. Even now when I see a family of deer in my neighborhood, I often pull over to take a picture to text to my Dad. Sometimes he'll text back, "Look at that pretty doe"—or point out some subtle detail I had not seen: "Look at the fuzz on his antler. He'll be a big one!" Other times, I'll get a silly text typical of my Dad, like: "Wish I was there with my gun!"

While I grew up deep in hunting culture, my Dad does not come from a line of hunters. He explains that, "No one in my family hunted or fished, except my Grand-Daddy Seal, who was an avid fisherman. In fact, when he died, I got all his fishing equipment. Most of my family (the men at least) were WWII and Korean vets and never had a desire to own a rifle or handgun." He reckons that's because those men saw rifles as weapons for war rather than sport. His Dad never hunted, and at the age of 15, while working in a local grocery store, the owner, Mr. Leo Strickler, came in with "a real nice British 303 rifle for sale." My Dad asked, "How much? And he said $25." Making only 45 cents an hour working after school and on Saturdays, it took him three weeks to pay Mr. Strickler. My Dad finally had his first rifle, "along with three bullets—two more than Barney Fife!" His aunt and uncle would take him hunting, and he watched his Aunt Mary Jane shoot an eight-point buck on his very first outing.

My Dad explains that by age 16, he was hooked: "I hunted every waking moment when not at work or school, and the summers were filled with fishing on the river." My Dad fished for trout in the Hawksbill (good for fishing and Jeep driving!), as well as in Rockingham County at a place called "Bennie's Beach."

I remember learning to fish with him on the banks of the Shenandoah River, a tributary of the Potomac flowing northeast and named by George Washington in honor of a Native Indian tribe. It's a storied river with local folklore and high school bragging rights about fishing or water skiing, getting drunk or getting laid under one of the bridges, and deaths of locals getting caught in the undertow. It's remarkable enough to be the stuff of songs, and I can't hear *Oh Shenandoah* or John Denver's *Country Roads* without pausing. Even now, it's not until I pass over the south fork of the river on Route 340 that I feel truly at home.

My Dad and I would fish for trout and smallmouth bass, stopping at Riverside Market for a pint of nightcrawlers on our trips to the Shenandoah. My Dad would usually grab a six-pack of Schlitz and a bag of ice. He'd buy me a bottle of Coke and brag on me to the old fella behind the counter about how I'd get my own worm on the hook (Fig. 9.2).

Fig. 9.2 Fishing on the river (1978); Dad teaching my son to fish (2016)

Other times on the river I didn't have my father's attention. My parents' friends, Dick and Connie Sours, owned land on the river, marked by a series of trailers where we would run free and unsupervised. By day, my Dad and the other men would go out on the boats and come back drunk, the trout taking their place among the empty beer cans in the cooler. After skinning and prepping the fish, my Mom would help the other wives dunk them in egg and cracker meal to fry on the grill. Forgotten by the grown-ups, we took turns jumping off the dock and competing for height and distance on the tire swing that flung us into the dark water. Save one, those summer friendships of mine were anchored to the river—and never really extended to school beyond awkward hellos in the hallway. Some of the kids lived in other, poorer parts of the county—a few even in the hollers. But none of that mattered on the river. It didn't matter that I was a straight A student or a bookworm, it was one's ability to slalom ski or flip off the tire swing that earned respect. I learned to ski on that river, summer after summer—the kids banished at night to the trailer farthest away while the parents drank liquor or beer and played cards long after we had stopped eavesdropping and sneaking our own alcohol (and on a brave night moonshine). I can still see my parents' friends smoking and talking, my Mom and Dad among them, forever laughing and looking younger by the water's edge.

Whereas my adolescence included these rites of passage, my father's was taken by Vietnam. At age 18, he volunteered for the draft and joined the Army. He explains, "This gave me even more experience with a rifle and handgun." After six weeks of basic training and another six of infantry training, my father went to Vietnam where his rifle would become the most important item to carry ("24 hours a day"). He added, "Thank God I never had to shoot anyone or use it to defend myself." Shortly after returning home from Vietnam in August, 1967, hunting season began, and "I was even more anxious to get in the woods." Like the river for me, I think being in the woods has always signaled home for my Dad. He still hadn't shot anything other than squirrels and rabbits and was still hunting for his first buck or bear.

My Dad would soon learn to balance his infatuation with another love interest: my mother. "When I asked my girlfriend's hand in marriage, I told her hunting was my love." I have watched the two of them, now married for 51 years, negotiate what it means for my mother to be a "hunter's widow" for three months of every year. The week before Thanksgiving, for example, my parents would act out their annual argument: my mother standing in the carport with her hands on her hips, shaking her head in judgment. She would say something to the effect, *I can't believe you're going at the holidays.* Meanwhile she would pack his cooler full of ham and cloth bologna, mumbling under her breath, *Well, I can't let him starve.* Burdened with guilt, my father would surrender, saying, *Fine! I won't go!* and then welcoming her anticipated response of, *Oh, just go!* At this, there was a quick exit with him driving off before she could change her mind, her calling after him to please be careful. This was before the days of cell phones so it meant a week of worry for my Mom.

Another time my Dad came home midweek to find my mother dealing with a stopped up sewer. "Of course, hunting took priority," my Dad explained, "so I got a 5 gallon bucket out of the building and put a toilet seat on it." My mother was not

Fig. 9.3 My Dad at Cedar Creek

amused: "Needless to say that didn't go over too well." And despite my Dad's proclaiming that hunting came first, he stayed home the next day to repair it.

Now 73 and retired, my Dad spends nearly every day between October 1 and January 4 getting as close to the mountains as he can. In fact, the week before hunting season, I can see the youthful 16 year old in him, that boy still full of wonder and anticipation (Fig. 9.3).

Cedar Creek: *The Place for Meaning Making*

That wonder manifests now in a retired man who can barely sit still the week before he heads to the cabin at Cedar Creek, a privately owned property adjacent to the George Washington National Forest. Weeks before hunting season opens, he and his lifelong hunting buddies meet in the mountains at a cabin they have taken care of for nearly 46 years (passed on through a tradition of seniority). They busy themselves with the tasks of fixing up the cabin, cutting firewood, and repairing deer stands. Meanwhile, my Dad's preparations carry on at home when he's not at the cabin— waxing his bow, cleaning his guns, and getting supplies—activities that are

Fig. 9.4 The woodpile at
Cedar Creek

exclusively his. In addition to taking care of the land and cabin, the hunters each pay
dues of $250 per year to cover the tax on the property (Fig. 9.4).

In 1974, my Dad met his little sister's boyfriend, Walt Keyton. Walt had grown
up in Manassas, then a distant suburb of Washington, D.C., but now a bustling bedroom community of Northern Virginia. My Dad explains that Walt had "grown up
with some guys in Manassas whose family had land in Shenandoah County."
Shenandoah County sits northwest of our county, Page, on the West Virginia line. As
my father's friendship with Walt grew, so did his relationship with Cedar Creek:

> My first trip to Cedar Creek on the Brill property was 45 years ago. Around 18 guys occupy-
> ing a three-room cabin that we still use. Many improvements have been made over the
> years, including electric lights and satellite TV. We used to walk over a mile in the mornings
> to get to our ground blinds. Now we ride 4-wheelers to nice tree stands.

And those tree stands all have names! My Dad laughed as he recounted the names
and their meaning:

- "Telephone Booth because that's the only place we get cell reception;
- Willis' Stand was built by Willis Jenkins but he hasn't hunted there for 20 years;
- Poplar Tree—everyone has killed a deer out of there. The tree is gone now but
  the stand remains;

Fig. 9.5 Uncle Walt on a
4-wheeler beside the cabin
at Cedar Creek

- Trouble Stand because it's too close to the edge of the property;
- Nate's Stand—he's the new guy who started coming two years ago;
- Skyscraper is Ernie's stand (there's also 10-Point Stand where Ernie shot a 10-point buck);
- Then we have 'The Fingers'—the First Finger is the ground blind; John always goes to the Middle Finger (he laughs);
- Mine is Square Stand because it's the size of a sheet of plywood." (Fig. 9.5)

One year, as my Dad explained, they made a camp rule: eight pointers or bigger were the only deer they could shoot. It was early on Thanksgiving morning when they decided to amend the rule: "Ernie, John, and I were the only ones at the cabin. We had agreed that we could break the rule for one day only." It was about 8 in the morning, "and I see a huge eight-pointer from the Square Stand." What he didn't realize, however, was that he was seeing one deer standing in front of another. "So when I shot, one deer fell and one ran." When he got to the deer, my father saw four nice points on one side of his head, but when he picked his head up, "all four on the other side were broke off. So was it a four or eight? I say eight." He still has the rack at home.

My Dad knows every acre of Cedar Creek, the tree stands, the cabin itself. They also have a shooting lane for target practice: "It's 123 yards to hit the bowling pin, and I've hit it with every rifled I've owned." My Dad talks about Cedar Creek in the way others talk about their world travels—with awe and adoration. It was a very special occasion when he took my sister and me to the cabin—only a handful of times and usually to go fishing. Though once I went squirrel hunting with my Dad, and he swears I loved it when he would skin them after the kill. As he tells it, "You loved watching Daddy 'pull the coats off' the squirrels." In the off-season, my Dad takes my Mom on trips to the cabin where they ride ATVs and hike the property. While my Dad shared Cedar Creek with us, they were isolated and rare trips. It was not a place we visited casually. Boyfriends and husbands were never invited. My Dad has never taken any of his five grandchildren. Cedar Creek is hallowed ground—and my own memories of it are from decades of hearing stories about "the cabin" rather than my personal experiences there.

Perhaps part of what makes Cedar Creek so special is the fact that he killed his very first deer there, "a nice six-point buck," and "I have killed a deer every year, except one, for 46 years, and the 'buck fever' is still there!" However, my Dad says that the most important thing about the cabin and hunting over the years has been the friendships he has fostered with men from around the state. The original owners are deceased and their descendants now own the property. My Dad and Walt are the oldest living of the original 18 and "have earned the right to be called 'seniors.'" My Dad says the friends are too numerous to mention but have meant so much to him and have been friends "through thick and thin." My Dad went on to talk about the time he fell and broke his leg at "the Crossroads." He was at the top of the mountain where, as a rule, no one goes alone. He got tripped up in some greenbrier while trying to move a log blocking a passageway. He fell like a tower and knew at once that something was terribly wrong. His buddies kept him calm and carried him out—not an easy task in that terrain. He added, "Most of these guys are only there a couple of weeks a year, but the friendships last forever."

Friends from the cabin "come from all walks of life, but hunting friends have one common bond: We all love the great outdoors and we all love to hunt." However, it's clear to me that those bonds go far beyond just the great outdoors. At my grandfather's funeral over a decade ago, I remember meeting a handful of strangers who had come to pay their respects. It was an emotional day, but I saw my Dad break a bit harder when these men entered the room. I imagine a lifetime of telling stories at the cabin meant these hunting buddies knew how difficult a day it would be for my Dad to bury his father. My Dad was so excited to introduce us to them and said things like, *This is Ernie. He's the one who… and here's Fred, he's the guy who works with slate. And you remember Pietro* (Figs. 9.6 and 9.7).

Not all of those visitors were strangers—a handful (like my Uncle Walt) are local to Page County. My Dad would visit with local hunting friends regularly for coffee or to take a quick trip to the woods for off season hunting, like spring gobbler season ("where you can only kill a gobbler with a beard") or to hunt morels, except we called them "merkels," derived from the Appalachian pronunciation of miracle. One such friend and perhaps his closest hunting buddy, Rondal Painter, passed away during the writing of this chapter. My father honored his friend the best way he

Fig. 9.6 The guys at Cedar Creek, early 1980s

knew how—he took to the woods on the Saturday evening after his funeral to go hunting.

In addition to losing friends, growing older, and watching a new generation of hunters use the cabin, my father hates to admit that times have indeed changed at Cedar Creek: "Nowadays, we do more complaining and playing cards than hunting, but I still love it there." My Dad used to skin and butcher deer at our house. Coming home from school in early winter to find a gutted deer strung up to the ceiling of our carport was an annual occurrence. But, just as the cabin and the men inside have grown up, so too has the process of cleaning deer. Now my Dad takes his deer to a local processor to have the meat packaged. He gets plenty of jerky to share, and we all look forward to my Mom's deer chili. He donates what we don't need to a charity called Hunters for the Hungry (Fig. 9.8).

I grew up on my Dad's bounty. It was never about bravado. He has no mounted or taxidermied trophies—just a few antlers displayed on a shelf at home. Plus, he says, "It's so expensive and the biggest one was only an 8-pointer." We had deer meat every which way you can imagine, chili, pepper steak, hamburger pie, meat-loaf, but it wasn't until college that I learned it had a more formal name. I was at a restaurant with a group of friends and saw an unfamiliar word on the menu. After the server responded to my question about the meaning of "venison," I said, "Oh! We call that deer meat in the country." Everyone at the table laughed, and said things like, "You're adorable."

Fig. 9.7 Ernie, Uncle
Walt, Dad, and Fred at
Cedar Creek, 2016

Fig. 9.8 Mom's meatloaf
recipe

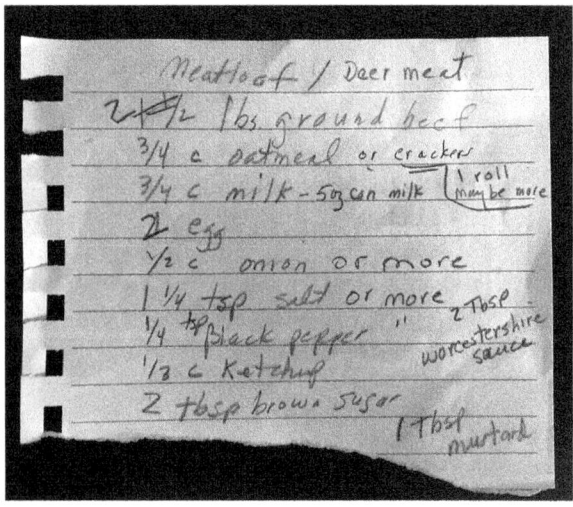

A Place Is More Than One Place: *A Critical Theme*

Deer—or *venison* rather—sustained us in the winter months, which is why my becoming a vegetarian in college seemed like a personal affront to my father. It was one of the first times (though certainly not the last) where my values started to diverge from my family's. While my Nanny often teased that I was getting "too big for my britches," it was clear that being outside of my small, rural community changed my relationship with it, whether I wanted it to or not. Regretfully, I lacked the language or the critical thinking skills to understand the complexities of my sense of place—and how the world (and my experiences of being called a hick and redneck) was challenging it. So, like many who leave their rural homes or small towns, I did so with good riddance. I have often wondered if my scholarship in rural education is my making amends—"prodigal daughter" of the Shenandoah.

My own fractured or splintering of place came both within and outside of my family. From those summers on the river, my closest friend was my Mom's best friend's daughter. We spent summers and holidays together, and when we were toddlers my Mom kept us both until we started kindergarten. I attended an elementary school on one end of the county, and she was on the other, poorer end. From there our lives diverged. She was loved at home, as I was—but our experiences took us in different directions. I left for college, and she stayed in the community and would work a series of minimum wage jobs. Eventually she served time in a federal prison for drug trafficking. I saw her recently for my Nanny's funeral, and we hugged and cried and laughed like old times. As we got to talking about our lives and children, she asked about my job. I told her I research rural education, and she laughed and said, "You need to fucking interview me."

Her *place*, much like my Dad's, was a very different one than mine—even though we all grew up in the exact same spot. Same mountains. Same river. I wonder how a critical lens for understanding place might have made us stewards of our own experiences. Had we been able to celebrate and critique our community, perhaps we would have felt empowered rather than, in some cases, limited by it. Being a vegetarian, for example, was something my Dad couldn't understand—why anyone would intentionally not eat meat seemed to him indulgent, a luxury of a class or generation of which he was not a part. My Dad grew up poor with just one of everything—one pair of shoes, one pair of pants, one set of toys. He was taught to eat everything on his plate and certainly never had a choice about what was on it. Out of necessity, he learned how to take exceptional care of his things. This made him into a particular man and even in our leanest years my Dad always had a clean car, a mowed lawn, and we sat at the dinner table until our plates were clean. These were my earliest memories of realizing how my place had shaped me in ways I have yet to fully understand, though I know my father's deep connection with hunting is emblematic of that culture and identity.

Hunting Literacies: *Ways of Being a Hyperliterate*

As a literacy professor, I can't help but zero in on the teaching opportunities learned from my Dad's experiences as an avid and enthusiastic hunter. I completed my dissertation in a rural school district in the foothills of Appalachia, and one of the first things I noticed pulling into the high school's parking lot was the number of gun racks on the cab window of students' pick-up trucks. There was plenty of camouflage in the hallways and, in the course of that study (see Azano 2011), I found that both male and female students alike talked about the importance of fishing, hunting, and riding four-wheelers—the same activities my father cherished then and now.

My Dad did well in school and excelled at math but as a "country kid" stayed mostly invisible. My guess is that he learned at an early age it was an expectation that he join the Army as the generations of men had done before him—not that post-secondary schooling would have been his choice or financially feasible. While the relevance of school might have been lacking, my Dad's work ethic certainly was not. In addition to working jobs at his home and for neighbors, he worked at Safeway after school and on weekends to save for his first rifle and also his first car. In a provocative piece of scholarship on rural education, Thomas Barone (1989) talks about the "Ways of Being at Risk" and describes a rural teen, Billy Charles Barnett, who excelled at place tasks such as hunting and fishing, yet struggled with school attendance and traditional academic requirements. In some ways, my Dad reminds me of Billy Charles not because my father struggled emotionally or academically, but because school offered no relevance to his real life happening outside the school building.

I asked my Dad once if he thought school would have been more engaging had he learned about biology through the animals he knew so well, geometry through archery, Earth Science through the lens of Page County, or English through texts about hunting and being outside. He just looked at me, and asked, *Is that possible?* As a philosophical underpinning to any curriculum, a place-based pedagogy isn't only possible, it could potentially be an academic lifesaver for rural students, a regenerative source of energy or sustainability in the health of rural places. If kids growing up in the country only learn about concepts or places that are remote and distant to their real life experiences, how will they find it relevant or engaging? Global competence has (understandably so) become increasingly important but should not come at the expense of understanding the local.

In thinking about the relationship between the school curriculum and a student's life, John Dewey (1897) suggested, "It is the business of the school to deepen and extend his sense of the values bound up in his home life" (p. 78). But how often do we see classrooms actualizing this principle? Rather, state standards (and as a result schools and classrooms) too often engage in what Paul Theobald (1997) calls the "decontextualized" curriculum. As a case in point, local to where I live (in Southwest Virginia), elementary age students—as part of their standardized Virginia curriculum—learn about the Nile River. My own children excitedly told me about one of the oldest and longest rivers, and knew a handful of miscellaneous facts about the river and its importance in Ancient Egypt. But when I asked them if they learned about another river that's also considered by some to be the oldest, they were confused. We live just miles from the New River, an ancient river that predates the Appalachian Mountains (believed by many scholars to be the oldest mountains in the world). These are the mountains and rivers we see daily in our local community; however, they are not prioritized in the curriculum. In my travels to many rural schools in these parts, teachers say they simply don't have time to add "extra stuff" into the curriculum—a sincere concern when they are evaluated by students' performance on standardized, state assessments. I often use the Nile and New rivers example as a missed place opportunity, but not simply as a way of making the curriculum more relevant but for allowing the type of critical inquiry needed to "reinhabit" a place—to "pursue the kind of social action that improves the social and ecological life of places, near and far, now and in the future" (Gruenewald 2003, p. 7). Learning about local waterways is indeed "place-based" but questioning why school curricula do not allow time for meaningful local contexts is, in effect, *critical*. Also, studying a place critically gives students permission to critique and develop a more nuanced understanding of the places they might value. Theobald (1997) suggests that place-based education can teach students about the "intradependence" between people and the environments in which they live, but when the curriculum is devoid of local community, those opportunities to foster intradependence, or to appreciate existing "within a place" (p. 7), are missed.

Rural school practitioners and scholars have long touted place-based practices as their bread and butter for locally responsive curricula. However, in my research, I have found that these pedagogical practices too often rely on any given teacher's belief that place matters. Meaning, in some classrooms, the content might be deeply contextualized. I've watched a rural teacher use country song lyrics to connect students with the teaching of *Hamlet*. But I have walked across the hallway to find a lesson packaged in a way that could be delivered in any classroom across the country.

I believe that in order to have a deeply contextualized curriculum, place-based practices must be adopted at a school or division level. A school or system-wide philosophical underpinning of curricula ensures that students are making locally relevant connections throughout the school year and that their understandings of place are deepened across the curriculum—and not simply to affirm place but to understand it critically as well. To question it. To challenge assumptions. To allow students the opportunities to think of themselves as active participants in their local community and what that participation means on a global level.

My father's connection to nature developed at a young age, and as a hunter he is part of a rich discourse community and culture. A critical pedagogy of place (Gruenewald 2003) might suggest that this rich place experience could be used to cultivate meaningful, instructional practices for a student like my father, and that these practices could allow for connections with social *spaces*, as a way not only to experience "the places outside of school—as part of the school curriculum" (p. 9) but also to consider what it means to live well in these spaces. Rather, school existed in a vacuum. It had little to do with my Dad's real life and interests. In some rural communities, schools are closed the first day of deer season. Why not use the significance of a local tradition to read (or write) poems about nature and deepen students' understandings of farming and sustainability, the ethics of hunting, politics of gun rights and the Second Amendment, and so on? Why not ask how relationships to social environments present opportunities to teach empathy and promote social action?

My Dad talks about hunting as a science that "takes many years to develop and perfect." Embedded in his hobby, my Dad was able to name many lessons he learned—all of which have implications for K-12 classrooms. As he explains it, to become a successful hunter, one must have an understanding of climate and weather conditions; geographies, travel, and bedding areas; game laws in your community; mating season ("when bucks come into rut"); and an intimate understanding of the weapon of use—bow, rifle, or muzzle loader. I would argue these historical and STEM-related concepts have critical, instructional implications for the classroom.

As a former English teacher, I can think of many missed opportunities to capitalize on some of these lessons in the humanities—Atticus picking up that rifle to shoot a rabid dog in *To Kill a Mockingbird*, the painful choice George confronted in *Of Mice and Men*, the place references in *The Yearling*, and poems like "The Farm" by Joyce Sutphen, "Taking to the Woods" by Henry Taylor, "What Cows Know" by Susan Blackaby, "Fishing" by A.E. Stallings, and my personal favorite, "Where I'm From," by George Ella Lyon.

Imagine the teaching of *To Kill a Mockingbird* as an opportunity to explore not only issues of race and social justice but also the positionality of rural, the way that poverty is leveraged as a rural trope in the novel (that is: the greatest poverty equates to the most despicable characters—as seen when comparing the Finches and Cunninghams and Ewells). A critical frame could serve to bridge understanding among the varied themes in the novel, or to forge new understandings as they relate to contemporary topics, such as the convergence or conflation of racial tensions and rural poverty. Using a critical pedagogy of place not as an instructional "strategy" to make the curriculum more relevant or engaging but as a theoretical underpinning to practice *conscientização* (Freire 1970/2009), to be conscious of the sociocultural and political relationship with the economics and ecology of a given place—this is the real possibility of place: for students to become, to become aware, to look at the past as a "means of understanding more clearly what and who they are so that they can more wisely build the future" (p. 84).

One Sight: *Resisting the Blindspot*

Hunting continues to be the fifth season in our family. I am completing the first draft of this chapter on the dawn of Thanksgiving where meals are still being planned around my Dad's need to get back to the mountains before dark. He will leave shortly after we finish our slices of pumpkin pie. My parents no longer argue about his departure. I think my Mom has grown to appreciate what his love for hunting gives him. Those lessons learned? My Dad says:

> I have learned so much about myself while in a tree stand in the woods. I have learned about patience, in life, even though I don't always show it. I have learned to be persistent and not give up. I have learned that love of family is always first. I can only hope that my health will allow me many more years of doing what I love to do.

During the writing of this chapter, it became clear that my Dad is still actively constructing meaning from his experiences at Cedar Creek. After the election of Donald Trump, my parents and I got into a conversation about gay rights over lunch (an issue of great importance in our family). My Dad got choked up and with deep regret said, "Like everybody back then, we used to call homosexuals fags or queers." He paused to wipe the tears from his eyes and confessed: "One night we were up at the cabin carrying on," when the most senior owner of Cedar Creek, Granville Brill, chastised the men for their conversation: "Granville's exact words were 'Love is the greatest gift we have, so if a man loves a man or a woman loves another woman, who are we to say it's wrong?' It sure taught all of us guys a wonderful lesson." Indeed it was a wonderful lesson for my father to learn; however, I have to believe that my Dad was willing to learn it because it came from a man whom he admired and respected—a hunting mentor. Granville Brill is a purveyor of place in my Dad's memory (as my father is in mine). That wisdom coming from someone outside of my father's discourse community might have fallen on deaf ears. (Imagine a school full of teachers enacting that same critical place pedagogy with students!) While my Dad says this particular experience was one of the "greatest lessons of his life," I would add that his ability to reflect and admit error is the greatest lesson I have learned by his example.

As I reflect on what I have learned about my father from this project, I hope that this work shines light on the type of rurality that Bill Green (2013) referenced as a bit of a "blindspot" (p. 26) and gives voice to the ways that unique geographies provide meaning for contextualized rural literacies. My father is a different type of "hyperliterate." His rural literacies grant access to a group membership, with hunting serving as the communal activity in which members make meaning from texts.

"Place-conscious educators contend that the most powerful forms of learning engage people in real problems and issues, beginning in their particular, nonstandard communities of practice" (Green and Corbett 2013, p. 3). I wonder if we perpetuate ways of being at risk (Barone 1989) when we don't value *place* and the home knowledge students bring to the classroom. I love to imagine a school day where, for a kid like my Dad, what a student values is reflected in the classroom, where "what I have to do" is met by "what I love to do." In this reflection, I am reminded of the last stanza of Robert Frost's (1934) poem, "Two Tramps in Mud Time," in which the speaker conveys a simple goal: "My object in living is to unite / My avocation and my vocation / As my two eyes make one in sight." Place-based education has the potential to provide that vision. Schools can be a site for these critical place understandings, where the next generation of rural students could learn that their rural literacies are in fact sustainable practices, both as a means for thinking of themselves as literate beings and also for making meaning in places they value. I know that would have made a world of difference to the inspired and ambitious young hunter in my father, Donnie Price. Why not school, indeed.

References

Azano, A. (2011). The possibility of place: One teacher's use of place-based instruction for English students in a rural high school. *Journal of Research in Rural Education, 26*(10).

Barone, T. (1989). Ways of being at risk: The case of Billy Charles Barnett. *Phi Delta Kappan, 71*, 147–151.

Dewey, J. (1897, January 16). My pedagogic creed. *The School Journal, LIV*(3), 77–80.

Donehower, K. (2013). Why not school? Rural literacies and the continual choice to stay. In *Rethinking rural literacies* (pp. 35–52). New York: Palgrave. https://doi.org/10.1057/9781137275493.0007.

Eppley, K. (2013). My roots dig deep: Literacy practices as mirrors of traditional, modern, and postmodern ruralities. In *Rethinking rural literacies* (pp. 75–92). New York: Palgrave. https://doi.org/10.1057/9781137275493_5.

Freire, P. (2009). *Pedagogy of the oppressed.* New York: Continuum. (Original work published 1970).

Frost, R. (1934). *Two tramps in mud time.* Retrieved from http://www.etymonline.com/poems/tramps.htm

Green, B. (2013). Literacy, rurality, education: A partial mapping. In B. Green & M. Corbett (Eds.), *Rethinking rural literacies* (pp. 17–34). New York: Palgrave. https://doi.org/10.1057/9781137275493_1.

Green, B., & Corbett, M. (2013). Rural education and literacies: An introduction. In B. Green & M. Corbett (Eds.), *Rethinking rural literacies* (pp. 1–13). New York: Palgrave. https://doi.org/10.1057/9781137275493_1.

Gruenewald, D. (2003). The best of both worlds: Critical pedagogy of place. *Educational Researcher, 32*, 3–12. https://doi.org/10.3102/0013189x032004003.

Theobald, P. (1997). *Teaching the commons: Place, pride, and the renewal of community.* Boulder: Westview. https://doi.org/10.4324/9780429496950.

Dr. Amy Price Azano is an Associate Professor of Adolescent Literacy in the School of Education at Virginia Tech. Her scholarship focuses on critical literacies, place-based pedagogy, and rural education.

Chapter 10
Gathering Sap

Jonathon W. Schramm

It's not every day that you realize you owe your life to a tree.

And not just in the general "all life is one" sense, but in tangible, specific ways. In the tender mushrooms that roil about in the simmering stew, and in the sweet syrup that rolls off of a morning's pancakes. These foods will soon make up a small but real portion of the molecules in my body. But equally so in the beauty of gently green leaves in the first blush of spring, and in the golden glow of leaves on a late afternoon in the fall. These gorgeous tones will soon take up their places in my memory, joining others linked to this tree in my experience.

"So how much sap do you need in order to get enough syrup?" asks my neighbor and friend.

"The old rule of thumb is 40 gallons of sap for one of syrup, but here in the city, where the trees grow broad without much competition, I usually find that I need about 32 gallons of sap for one of syrup."

"But how do you actually make the syrup?" persists his 6-year old son, focused on more important questions than sap-to-syrup ratios.

The best answer to that is to see it happening, so we walk from their yard where I have hung a galvanized tin bucket on one of their trees, to my backyard deck, where a cloud of steam rises off of a broad metal pan stationed over a propane burner.

"You need a fire hot enough to boil the sap, ideally as fast and efficiently as possible, so that the sap becomes more and more concentrated. The sugars in the sap will caramelize – getting even sweeter and more interesting! – as they boil. I use this broad dish so that there is plenty of area for the steam to evaporate off."

"But is it ready after you do all that?" asks the dad.

"Close, but I will take it inside once it is a good bit thicker and a light brown color, which ends up being a pint or so of liquid from the initial 5 gallons in the pan. Then I finish evaporating off the last bits of water on the stovetop." (Fig. 10.1)

J. W. Schramm (✉)
Sustainability and Environmental Education Department, Goshen College, Goshen, IN, USA
e-mail: jschramm@goshen.edu

© Springer Nature Switzerland AG 2020
J. B. Pontius et al. (eds.), *Place-based Learning for the Plate*, Environmental Discourses in Science Education 6,
https://doi.org/10.1007/978-3-030-42814-3_10

Fig. 10.1 A simple back porch evaporating set-up. Essential are a heat source (propane burner in this case) and a broad tray from which sap can evaporate, allowing the sugars to concentrate. Extremely helpful are components to minimize wind's effect on your heat source (the sheet metal skirt) and an adjustable lid that lets you bring a fresh batch up to boiling as efficiently as possible (the baking sheet)

Maple syruping is, for me, an annual rite that reaches back to my childhood and I hope, will stretch ahead into my children's lives. Coming as it does at the very end of winter – or the very beginning of spring, depending on your perspective – it represents a tremendous miracle of life springing forth even while snow and ice are all around. Although my small home scale of tapping – usually six trees with one bucket each – is modest compared to the rural sugar bush that I grew up working in with old farmers from my home church, the natural phenomenon that both rely on is the same. And so are the pleasures of smell and taste that result from the work.

I still enjoy the style of collection that I grew up with, galvanized aluminum spiles and collecting buckets on the side of the trees, each with a peaked lid to keep out snow and rain. Because they are so conspicuous, these buckets are conversation starters when people see me maintaining them in my neighborhood. One year, as an elementary school kid asked me, "Is that a squirrel house that you put up?" Not an unreasonable question, I suppose, and he was probably not the only person wondering that. I usually rotate collecting from different sugar maples in the neighborhood, which has given me opportunities to meet many of my neighbors. Appearing on their doorstep in a season during which few people spend much time outdoors, my request to tap their trees seems to catch them off-guard, but after a little explanation, most have been quite content to let me start to work. And once they get their share of the sweet produce at the end of the season, there is hardly hesitation the next time I ask (Fig. 10.2).

Fig. 10.2 A galvanized aluminum bucket and lid to gather and store the running sap for a day or two. Both rest on the spile, a small metal spout that is tapped into the tree at the start of the season. Selection and drilling of the hole for the spile is crucial, as not all locations will produce at the same rates, depending on the health of the tissue. Most years the sap will run out of a hole for about 4 weeks before the tree closes the wound. Starting the season too early runs the risk of too many days not getting above freezing, thus keeping the sap trapped. Start too late and you may not have enough cold nights well below freezing that help the sap run to reset and remain a strong flow. Climate change is making this a more difficult prospect across the range of *Acer saccharum*, as the late winter period becomes much more variable. One response I am experimenting with is staggering the start dates of my various taps, so that some may always be in peak production across the season

Maple syrup is a surprising food: a fragrant, caramelized sweet that is drawn from the bark of living trees. And this miraculous product is also a wonderful way to show people that even familiar creatures around us (my hometown names itself "The Maple City" – sugar maples are abundant on our streets) can do amazing things, including providing for us, their human neighbors. It is a low-tech way to make a small amount of your food, and is a great way to begin becoming more place-based in your awareness and diet.

> *"By finishing the process inside, I can make sure I have the sugar concentration just right. If the syrup is too watery, it will spoil in storage, if too dense, it will crystallize in the jar and be much harder to use. A hydrometer helps me tell that I have the density right."*
> *"A hydrometer? What's that?" exclaims his son.*
> *I step inside the house and bring out the metal tube and glass instrument that help me gauge the density of the syrup.*

"It's basically a precise glass tube with a weight in the bottom, that can be placed in this metal tube with some syrup in it. It sinks down in the liquid, but when the syrup is just the right thickness, it will make the glass tube float at a height that matches this line. That's how I know it's not too thick or too thin."

"Finally, right before I can the syrup into jars, I use a felt sheet to strain the syrup and collect all of the 'sugar sand.' These crystals of minerals from the sap can really make the syrup unpalatable, and so I want to get them out if possible. It's important to do that filtering while the syrup is still hot though, or it will never run through the felt!"

"That's a lot more work than I expected it would be!" exclaims the son.

"True, but it's so worth it when you have homemade syrup on your pancakes in the fall, or on your vanilla ice cream in the summer!" (Pro tip: real maple syrup on a quality vanilla ice cream is most excellent eating!) (Fig. 10.3)

I harvested maple syrup in this way for some years contentedly, mostly just enjoying the process and the product, but eventually I started to become more conscious of the variability in sap run from year to year and from tree to tree. How the placement of a tapping spile – whether on a prominent arm of the tree or a quiet divot, under active, healthy tissue or a slowly senescing one – could sometimes dramatically affect the amount of sap that would flow from that tap. Or how the exact rhythms of freeze-thaw – or more often under a warming climate, durations of above-freezing weather – influenced the overall yield in a year. Or how the amount of precipitation that fell in the previous fall and winter seemed in large part to determine the amount of sap the trees had available. I have yet to come across any

Fig. 10.3 Finishing set-up within a home kitchen. A large pot is needed to act as the final evaporator, with the metal basket holding the felt filter which represents nearly the end of the process (only canning remains). Another essential tool is the glass hydrometer and metal tube, whose function is explained in the text

formal research confirming this last point, but I'm quite confident in the correlation I've seen at my house, and the physiology of the plants makes it plausible to me.

My growing awareness of these finer variations, rather than leading to a sense of mastery of the process, has only increased my sense of awe at the whole relationship. I'm more aware of both the strength and the frailty of the trees I draw from, acutely conscious of their resilience but also of the stressors that they can experience over the years. And, as happens anytime that you live around plants long enough, there is no better place to reflect on this than when they die, or at least when parts of them do. Several years back I had to trim a few limbs from my main backyard maple, and then wondered if I might find some way to use these limbs too as a gift of the maple.

The high whine of my electric drill cut through the growing darkness, firing in short bursts as I drilled hundreds of holes in a short time. I knew I had started the work too late when one of my neighbors came over and gently requested that I stop for the night, as the repetitive noise was keeping her kids awake. Wrapping up the log I was working on, I shifted to the quieter work of inoculating the logs with sawdust mixed with spawn of several edible mushrooms I was trying to cultivate. This simply involves filling each hole with a clump of inoculated sawdust – and here a handheld inoculation tool is a huge help – then sealing the top off with a bit of wax to keep the fungal spawn from drying out before it gets established in the log. As with making the syrup, inoculating these maple logs is not complicated, but requires some patience and a little elbow grease, along with a willingness to work with what the tree provides.

The next day another neighbor came over to see what all the drilling had been about the previous night, and I had the chance to explain to him a bit of the process.

"I started by separating the different parts of the tree branches by size, with basically three size categories: large branches a foot or more in diameter, medium branches from about 4 inches to 1 foot in diameter, and all of the smaller twigs and branches. This last I made into wood chips with a rented chipper, and spread in a thick mulch layer around the base of the maple. Into this I spread chunks of the sawdust spawn for a quick growing species, the winecap. I expect they'll be able to digest the wood chips over the summer and yield mushrooms already by next fall, and certainly by the following spring. I used the drill-and-fill method on the medium-sized branches with several types of oyster and oysterling mushrooms being the intended species from them. The pieces are bigger and it should take the mushrooms longer to come to fruition, maybe 12–24 months. Finally, the largest pieces will get special treatment. I'll slice them into thick cross-sections, and layer spawn of lions-mane and comb-tooth mushrooms in between, then store them in my cool basement for a year in thick garbage bags before bringing them back to the light of day next summer, where they will finish maturing and eventually fruit after 2 or 3 years!"

"Wow, that's a long time frame," responded my neighbor, "Is it because the wood pieces are so thick and dense?"

"Exactly," I answered, "but those are the tastiest of them all, I think. The complex work of the fungi seems to result in equally complex flavors." (Fig. 10.4)

Over the last seven years that we have lived in our house, I have experienced deepening ties to this tree in both tangible and intangible ways. The tree has contributed food to my family's table and joy to our times outside. We've come to see this familiar creature as simultaneously wilder and more domestic than we first understood it. We've learned that we can rely on the tree and our other more-than-human neighbors to provide for us in real ways, even as we've become more attuned to the activities and rhythms of these neighbors. This growing attentiveness inevitably

Fig. 10.4 (a) The "drill and fill" method in action. Simple tools available from several online suppliers make the process much more efficient (Image from Field and Forest Products, Peshtigo, WI). (b) A few of the "first fruits" of the winecaps inoculated into the wood chip patch, fruiting after 4 months

helps us to better feel the wildness of these creatures, despite their apparent ease around our domestic settings. And although I still treasure times spent in wilder settings than my neighborhood in Goshen, through this tree I've come to understand that foraging, and its effects on our spirits, can happen wherever we are.

As an environmental educator, I've also found it quite easy to slip into a testifying or educating role with our human neighbors and friends about these connections to our maple. Although I haven't done any formal post-assessment – the better to maintain the bonds of neighborliness – I'm sure that these conversations have also helped our neighbors to better appreciate and enjoy these trees that shade our neighborhood. Carrying out such conversations on a broader scale might just be an important way to grow an environmental ethic in an increasingly-urbanized North American population.

   As I have matured, I've come to see that the greatest values of local, seasonal living and diet lie not in helping us to mark our time, but in helping us to *experience* our time – to feel in deeper ways the connections between beauty, awareness and sustenance. The passing of each year means not just the passing of four seasons, but the crashing of four distinct waves of life upon the shores of our consciousness. Each wave has stages, degrees, complexity and depth, and because there is always more to experience in each wave than we are able, we can look ahead to the next year's version of that wave with great anticipation and eagerness. The power of such a realization for sharpening our senses and our gratitude for life on this planet is something that I feel I am just beginning to explore.

Jonathon W. Schramm is a plant ecologist and environmental and sustainability educator. He is a professor at the Merry Lea Environmental Learning Center of Goshen College, in the glacial hills and lakes of northeastern Indiana. He is interested in seeing the plant diversity of both rural and urban landscapes be increased, and is convinced that an involved and interested citizenry is crucial for caring for the diversity of our places. He lives with his family in the city of Goshen, Indiana.

Chapter 11
Towards a Marine Socio-ecology of Learning in the South West of England

Alun Morgan, Emma Sheehan, Adam Rees, and Amy Cartwright, with Kieran Perree, John Walker, Neville Copperthwaite (Contributions)

Introduction

Contemporary society is recognising the need to move towards a more sustainable future. This is particularly true of coastal communities associated with the fishing industry. There has also been an increasing acknowledgement of limited attention to the sociological aspects of marine fisheries research (Urquhart et al. 2014b). We wish to extend this critique to acknowledge the even more limited attention to the learning dimension of fisheries research. This chapter seeks to start to redress this lack by exploring how attention to social learning can contribute towards a "sustainable development paradigm for fisheries ... [which] entails thinking more explicitly about ... community, wellbeing, identity, gender equality etc." (Urquhart et al. 2014a, p. 2). We adopt a socioecological systems approach (Poe et al. 2014) which attempts to dissolve binaries between: nature and culture; expert and layperson; scientific and traditional knowledge systems. Furthermore, we focus on 'Place' as an integrative concept. Such a 'place-based' orientation is becoming increasing apparent in sustainable fisheries policy such as with Area Based Development (Budzich-Tabor 2014); and highlights the relevance of 'sense of place' and

A. Morgan (✉)
Institute of Education, University of Plymouth, Plymouth, UK
e-mail: Alun.Morgan@plymouth.ac.uk

E. Sheehan · A. Rees · A. Cartwright
Marine Institute, University of Plymouth, Plymouth, UK
e-mail: emma.sheehan@plymouth.ac.uk; adam.rees@plymouth.ac.uk; amy.cartwright@plymouth.ac.uk

K. Perree · J. Walker
Lyme Regis, England

N. Copperthwaite
London, England

© Springer Nature Switzerland AG 2020

J. B. Pontius et al. (eds.), *Place-based Learning for the Plate*, Environmental Discourses in Science Education 6,
https://doi.org/10.1007/978-3-030-42814-3_11

'identity' for 'learning for sustainability' in the context of fishing communities (Garavito-Bermúdez and Lundholm 2017; Worster and Abrams 2007). Such an orientation is mirrored in the emerging field of Place Based Education (Gruenewald 2003; Morgan 2012). The fisheries sector, as an example of a primary extractive industry based on a close relationship to the natural world and particular coastal 'locales' and associated traditional knowledge systems, lends itself particularly well to these explorations.

Until recently much mainstream, policy research was based on a "top-down science-driven fisheries management model" (Dubois et al. 2013, p. 48). Fortunately, there has been a move towards a more socio-ecological perspective in which 'fishers' themselves are acknowledged as key stakeholders/experts (Mackinson and Wilson 2014). Evidence suggests that their involvement can precipitate a learning journey leading some observers to talk about the development of the 'scientific fisherman' [sic.] (Dubois et al. 2013). We propose that there is likely to be a corresponding *learning* shift in the knowledge base and identity of the scientists. If in evidence, these would represent significant instances of 'border crossing' (Giroux 2005) between knowledge constituencies.

Yet there are other equally significant antecedent 'learning' trajectories we wish to explore. We are interested in the processes by which an individual 'learns' to become a 'fisher' in the first place. This draws attention to the mentoring and socialisation processes undertaken through ecological experiential learning 'on the job'. A similar learning journey will be undertaken by those becoming marine scientists. Then there is the question of how one 'learns' to belong to the 'community of place', whether a fisher or not. This focuses attention on local traditions and sense of identity. Finally, there are questions about how intergenerational learning for sustainability might contribute to all of this. This gives rise to a complex 'marine socio-ecology of learning' which we wish to explore in the South West of England, and the Lyme Bay Marine Protected Area in particular which forms the particular focus of this chapter.

Social Learning Theory

This chapter draws extensively on 'social learning' theory, particularly as it relates to the development of 'communities of practice' (COP) (Lave and Wenger 1991) in the context of 'sustainable environmental management' (Keen et al. 2005b). Social learning "tends to refer to learning that takes place when divergent interests, norms, values and constructions of reality meet in an environment that is conducive to learning. This learning can take place at multiple levels i.e. at the level of the individual, at the level of a group or organisation or at the level of networks of actors and stakeholders" (Wals and van der Leij 2007, p. 18). Theorising in this field provides a number of tools and concepts for exploring the learning in evidence within the Lyme Bay Case Study. We explore this principally with respect to the two key constituencies involved in the project, namely, marine biologists and commercial fishers. More crucially, we explore social learning at the interstices of these two groups as a small group comprising members from each COP come together to form a new and emergent inter-professional COP in the context of research praxis. Furthermore, we are interested in the relationship between this emerging COP and the wider community, particularly in relation to 'community outreach' activities and the promotion of 'Ocean Literacy' (NOAA 2013).

Becoming a Professional/Member of a Community of Practice

At one level, we wish to explore the learning processes involved in people becoming members of each respective profession. This draws attention to social learning theory around professional learning in which 'novices' or 'newcomers' gain experience and, ultimately, acceptance and membership through the development of professionally-relevant competence. This learning often has a formal dimension (e.g. certificated courses whether in practical skills such as 'competent crew', 'safety at sea' and 'boat handling', or in academic study for BSc, MSc, PhD), but is most likely to occur in the more informal learning context of everyday work practices, that is 'situated learning' and knowledge (Lave and Wenger 1991) since learning "takes place through working practices" (Hildreth and Kimble 2004, p. x). Thus, "[n]ovices (or newcomers) are enculturated into the practices of the community of practice and move from being able to participate peripherally in the practices to being able to participate centrally when they are competent in the practices" (McCormick et al. 2011, p. 44). Furthermore, it is a process that is inherently social and intergenerational, being facilitated by the mentorship of existing experienced members. Novices then 'rise through the ranks' until they, in turn, become fully fledged professional in their own right – thereby moving from the peripheral to full participation (Lave and Wenger 1991) – and capable of acting as mentors to the next generations.

This model is easy to recognise in the commercial fishing profession where a novice is apprenticed to a seasoned skipper and gradually rises through the ranks or crew as their experience and competence accrues and their roles change, becoming 'old timers' and 'old hands' in their turn. Arguably, however, this is the same process at work within the 'marine biology' community where one has to 'rise through the academic ranks' from undergraduate through post-graduate and, ultimately, fully fledged academic-researcher. Expressed this way, learning to become a professional is a collective or social endeavour. However, "[l]earning is a trajectory of becoming, and this trajectory is an individual one, even where that identity is shared with a community" (McCormick et al. 2011, p. 49). There is a further apparent commonality between those who undertake a career as commercial fishers and those who become environmental, or more specifically in this instance, marine scientists; namely, individuals often express a strong sense of 'vocation' or 'devotion' to the work and associated lifestyle rather than being motivated by the mere pursuit of financial remuneration. This relates to Stebbins's (2009) notion of 'devotee work', in particular as it relates to that involving significant 'Nature Challenge Activities' (Davidson and Stebbins 2011) as a consequence of the setting for this work, namely the sea or ocean which can be both awesome and awful. Thus, we are interested to trace some biographical details at the individual level concerning choices made to enter their respective chosen profession; and the subsequent learning undertaken in their professional journey.

So much of what marks out a 'professional' is the knowledge, skills, values and norms they have learned and internalised as part of their professional identity, which they share with fellow members, and which is 'situated' within that particular professional context. Consequently, it is appropriate to talk about professions as 'knowledge communities', 'communities of practice' (COP) or 'traditions of understanding' (Brown et al. 2005b), terms which are equally applicable to 'marine biology' and 'commercial fishing'; and, by extension, Marine Science Communication. Another sociological concept that relates to the foregoing discussion is 'cultural' or 'social capital' (Bourdieu 1986; Eames 2005) which describes the non-economic resources that accrue from membership of a group (including a profession) which can support social cohesion amongst, and mobility of, its members. Essentially, entering a profession can be seen as a learning process by which one can 'tap into', and in turn contribute to, the existing cultural and social capital associated with that profession and rise within it. The processes by which members within a group develop social capital has been referred to as 'bonding social capital' (Putnam 2000). This

> … is formed by the links between members of comparatively homogenous subgroups (for instance farmers, teachers, conservationists and small business owners). Bonding social capital is underpinned by some agreed sharing of identity or purpose, acting within communities to bolster the ties between members, and reinforce loyalty and support within the group […] The ties may be choice of occupation [… such as fisher or marine biologist] or more deeply embedded factors such as ethnicity, religion and identity. The tendency is to encourage inward looking groups (Eames 2005, p. 84).

The related COP concept is very much connected to learning processes at both collective and individual levels since it presents a "model for the creation of new knowledge and meaning … [and] also provides a home for the identities of members through the engagement in the combination of new types of knowledge and the maintenance of a stored body of collective knowledge" (Hildreth and Kimble 2004, p. xiii). Generally, COPs are believed to share certain characteristics: mutual engagement; joint enterprise; shared practices; shared repertoire and resources; common background or shared common interest; common purpose/motivation; dynamic nature (evolve over time); formal and informal relationships; narration or storytelling for knowledge sharing and knowledge generation (Hildreth and Kimble 2004). It is also important to note that COPs (including professions) as do all communities "have *boundaries* [emphasis in original] between them, and these provide a focus for understanding practice (the notion of legitimate peripheral participation is premised on moving from a boundary of a community to full participation at its centre)" (McCormick et al. 2011, p. 47). Becoming socialised into a culture can be seen as a form of 'boundary work' (Riesch 2010) in the sense of developing cultural 'insidedness' and identification with(in) a particular culture vis-à-vis that which is outside the 'cultural boundary'. This is associated with adopting the 'bounded rationality' (Bäckstrand 2011) of that discipline/profession/COP or 'knowledge tradition'. Furthermore, the existing of an identifiable 'boundary' permits another type of 'boundary work', namely 'policing of the boundary' to ensure legitimate and exclusionary membership. This 'policing' of the border and 'border politics' has been discussed in relation to 'science' (Gieryn 1999).

Learning to 'Cross and Bridge Borders'

Whilst the same general learning processes and trajectory – apprenticeship-mentorship, internalisation (of knowledge, skills and norms – the bounded rationality), bonding social capital etc. – might arguably be shared between fishers and marine biologists, the nature of the profession or 'knowledge community' to which they ultimately subscribe are very different; and each will be largely inward looking. On the one hand, professional marine biologists represent a subset of the 'Western Modern Science' (WMS; Snively and Corsiglia 2000) 'knowledge tradition'. Commercial fishers on the other hand, are more likely to develop what has been termed 'traditional ecological knowledge' (TEK), 'practical experiential knowledge' (PEK) or 'folk science' (Blount 2003). Traditionally, each respective 'knowledge community' has independently and exclusively subscribed to its own internally coherent logics and procedures – its bounded rationality – without recourse to the other. Indeed, they are likely to employ their own strategies for 'policing the boundary' of their respective COPs; and antagonisms are likely to exist between them. However, as noted in the introduction, there is increasingly a recognition that these two knowledge constituencies could have much to learn from one another, and a more adequate understanding of the natural environment would bring together scientific and local-experiential ways of knowing (Mackinson and Wilson 2014). This is because different traditions of understanding – whether "related to a discipline, professional training or cultural norms of a locality. […] can serve as a network of prejudices or pre-understandings that provide possible answers and strategies for action. Traditions are not only ways to see and act, but also ways to conceal. When traditions of understanding differ, dialogues are needed to understand each person's perspective and negotiate a way forward" (Brown et al. 2005b, p. 248). Thus, neither marine scientists nor commercial fishers alone have the monopoly on truth about the marine environment. Indeed each has its own blindspots. However, there are significant barriers to achieving this desirable cross-COP dialogue on a number of levels.

First there are epistemological issues of how legitimate knowledge should be achieved (the universal 'scientific method' versus experiential place-based knowledge); then there are associated challenges of communicating across different knowledge traditions, each with its associated jargon and practices; and finally, but perhaps most significantly, there are political issues of power asymmetries since "scientific, technocratic and economic information and knowledge that I deemed to be fundamental to resolving environmental issues [… whereas the fishers' c]ontext-dependent knowledge is afforded very little status or is considered secondary to the generalizable knowledge that supports government and other centralized views of capacity building […]" (Andrew and Robottom 2005, p. 65). Thus, "[t]he positivist, scientific mode of inquiry underpinning the adaptive management framework has emphasized modelling, and gathering measurable and objective data on the impacts of specific actions […] A tension has emerged between this focus and that of local stakeholders, whose knowledge often stems from experience and detailed knowledge of the local context" (Keen and Mahanty 2005, p. 114). This often leads to mutual suspicion and 'border disputes' between knowledge constituencies (Gieryn 1999), giving rise to a significant intellectual, cultural and 'political' 'barriers' in coastal communities where there is often mutual suspicion and distrust between marine science researchers and commercial fishers (Blount 2003), resulting in exlusionary 'boundary work' and antagonistic actions. This is a serious impediment to the goal of acknowledging 'fishers' as key stakeholders/experts as called for in contemporary sustainable marine science discourse (Mackinson and Wilson 2014). Ideally, "Co-management involves a partnership between different stakeholders for an area or set of resources for which common goals have been established […]. Co-management recognizes the value of different types of knowledge (for example scientific and local) and integrates local and higher-level management systems" (Keen and Mahanty 2005, p. 105). This, however, requires the development of trust and communication.

Fortunately, social learning theory, particularly in relation to environmental management and sustainability, has been particularly exercised by these challenges since it is concerned with "the collective action and reflection that occurs among different individuals and groups as they work to improve the management of human and environmental interrelations" (Keen et al. 2005a, p. 4). Once again drawing on the work of Putnam (2000), workers have used the concept of 'bridging social capital' which describes the processes by which heterogenous groups can become connected in a looser affiliation. The benefits of bridging social capital are " … the building of networks, relationships and connections among [and between] the bonded groups, linking heterogeneous groups with shared interests. Consequently, bridging social capital breeds outward-looking networks and brings together people from a variety of backgrounds …" (Eames 2005, p. 84). Such would be the situation should, for example, members of the Fishing COP work collaboratively with members of the Marine Biology COP. This is what happens in the Lyme Bay Case Study.

In many ways, 'bridging social capital' demands the reversal of the learning process, or indeed an 'unlearning', associated with becoming a professional, since:

As part of professional education and experience, we are trained to internalize the boundaries of our field of practice and accept the ethical and social constraints on our roles [… whereas in] in social learning for sustainable environmental management, the goal is the opposite. Rather than working with discipline boundaries, they have to find ways to transcend them, or at least build bridges across the disciplinary and social divides … Social learning in environmental management will thus involve opening up boundaries and combining practices (Brown et al. 2005a, p. 225).

This introduces the possibility of a more progressive and inclusive form of 'boundary work' than mentioned earlier which "… entails questioning the *borders* between science and non-science, expert and lay knowledge, universal and local knowledge" (Bäckstrand 2011, p. 449 [emphasis added]). The end result is, ideally, "… the ability to go beyond the boundaries, whether they are between places, people, governments and/or knowledge sectors" (Brown et al. 2005a, p. 240). Such a position can lead to a more dialogic 'boundary work' and 'border crossing' between multiple 'bounded rationalities' or COPs. This opens up the possibility of considering, through dialogue and mutual learning, the relative complimentary or contributory 'expertises' of different groups on their own terms as a matter of 'cognitive justice' (Leach and Scoones 2005). Furthermore, these "dialogues across traditions of understanding are a time for learning and development […] through each stage of the learning cycle, the team members will need to bring together a portfolio of skills. Diagnosis of the situation needs keen and accurate observation, coupled with some experience of the people and place" (Brown et al. 2005b, pp. 248–249).

However, "[f]or co-learning to be successful, bridges among the diverse actors involved in learning processes need to be built. The bridges can be built by facilitators with skills to communicate across social groups, institutional structures that facilitate learning among knowledge communities, or learning processes that transcend disciplinary and social divides and create new tools and symbols" (Keen and Mahanty 2005, p. 117). In the Lyme Bay instance there were no such facilitators or institutional structures, but the participants managed to find strategies and tools to enable communication amongst themselves.

Brown and Pitcher (2005) adopt another metaphor (after Dening) which is powerful and fits nicely with a discussion of learning in the marine context – 'islands' and 'beaches' (although it is important to recognise that the metaphor is equally applicable to terrestrial and urban contexts since these symbols are 'cultural' rather than 'physical'). Thus, islands represent a particular, self-contained community; whilst the beach represents a (potential) meeting place for representatives from different islands to meet. The task is one of creating 'beachheads' for communication to take place. Arguably, this is what is happening in Lyme Bay – the fishers and scientists involved meeting on the metaphorical beach of the project and literally on the floating island of the fishing/research vessel.

Casting the Net Further – Community Outreach and Intergenerational Learning

The final area of social learning we wish to focus attention on is the wider community. This introduces another COP germane to the discussion – that of 'marine science/management communication and outreach'. Too often, the results of academic field research have remained enclosed within the academic community. Fortunately, there has been a recognisable shift towards 'Public Engagement' (House of Lords 2000). Indeed, this is now reflected in the nature of academic funding bids and research grants which are now required to demonstrate an 'impact' plan. This is often reflected in outreach activities. Arguably, it is also behind some of the emphasis on scientists working more closely with non-scientists (such as fishers). In broader educational work, there has also been a shift towards social learning that is 'place based'.

This 'place-based' focus also relates to an emergent interest in heritage generally, and in coastal communities particularly (Howard and Pinder 2003). The fishing heritage of many coastal communities features strongly in this (Acott and Urquhart 2014; Martindale 2014; Urquhart and Acott 2012), with a particular emphasis on traditional ways of working and associated folk traditions. Finally, there has been an emerging focus on Ocean Literacy (NOAA 2013). Much environmental education that is terrestrial and human-environment interactions are reasonably immediate and obvious. However, the marine environment is very challenging in terms of awareness raising since it represents an aspect of the 'hidden commons' where many of the issues are 'out of sight' and the purview of only a small number of people – fishers, marine scientists and recreational divers. Mainstream society is generally not aware, which greatly enhances the need for marine science outreach and education. Taken altogether, these various strands of social learning represent the complex social ecology of learning alluded to in the title. These might be suitably represented in the following diagram (Fig. 11.1).

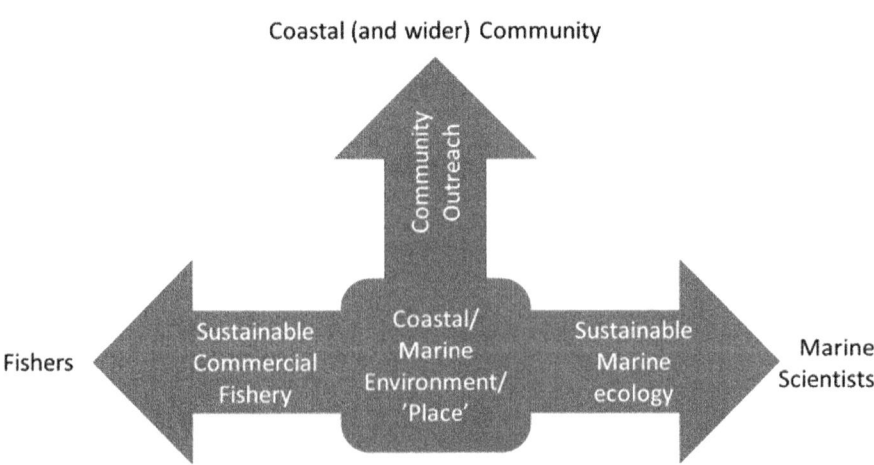

Fig. 11.1 Schematic of social learning dimensions and learning-practice constituencies

Lyme Bay as an Exemplar

The Study Area and Context

Lyme Bay is an extensive area of open water on the South Coast of England, UK encompassing parts of Dorset and Devon (see Fig. 11.2). This is an important area in the economy of the region in terms of both tourism and fishing. The offshore has a range of habitats, including rocky and cobble reefs, mixed pebbly sand and gravel sediments and muddy soft substrata (Sheehan et al. 2016). The area is also important in terms of biodiversity, holding many important Biodiversity Action Plan (BAP) species including seagrass (*Zostera marina)*, nationally rare sunset cup coral (*Leptopsammia pruvoti*) and pink sea fans (*Eunicella verrucosa*) – a cool water, soft coral (indeed, the study area is known as 'England's Coral Garden' in promotional literature). However, concerns were raised about the damage to marine reef habitats caused by dredging fishing practices. Consequently, from July 2008 the Department for Environment, Food and Rural Affairs (DEFRA),the UK governmental agency charged with environmental protection legislated for the closure of a 60 nm² area to bottom towed demersal fishing gear whilst still allowing sea angling, scuba diving, other recreational users and fishers using static gear such as pots and nets (The Lyme Bay Designated Area (Fishing Restrictions) Order 2008). In 2010 an extended area was proposed as a candidate Special Area of Conservation (cSAC) to include

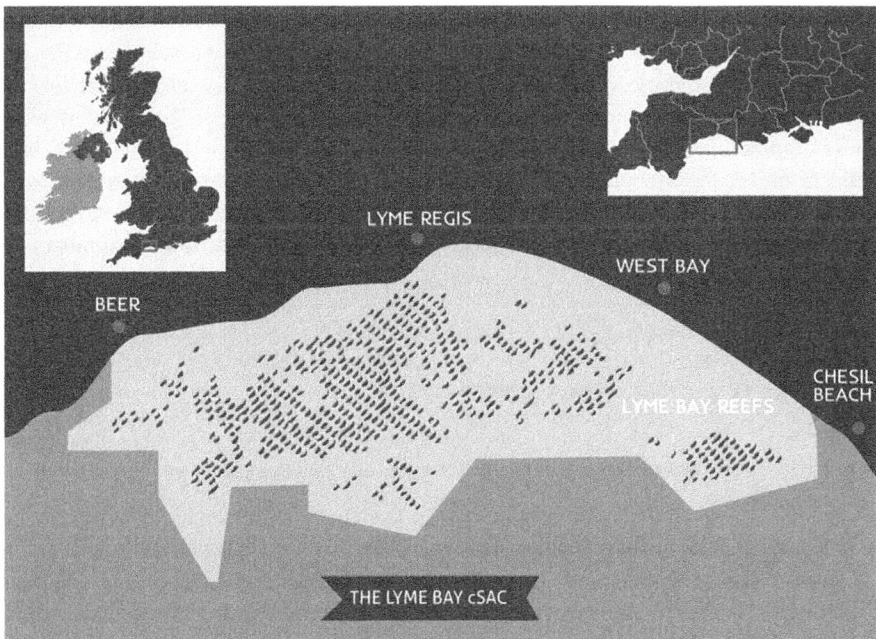

Fig. 11.2 Lyme Bay special conservation area – reproduced with permission from Blue Marine Foundation

important reef habitat which was accepted by the EU in 2011 as a Site of Community Interest (CSI) – referred to as the 'Lyme Bay Marine Reserve' (see Fig. 11.2). The principle aim of the closure was to maintain the structure of the reef system and to enable the recovery of the bottom living invertebrates (Sheehan et al. 2016). Since then a consortium of marine biologists led by the University of Plymouth (UoP) have undertaken annual surveys to assess the impact of this moratorium on dredging in the marine protected area, funding being provided first by DEFRA and subsequently by Natural England, an advisory body to the UK government, and the European Maritime Fisheries Fund (Sheehan et al. 2013). Surveys utilised High Definition Video cameras using either a towed 'flying array' (200 m video transects) or Baited remote underwater video (BRUV) to capture evidence of both sessile and sedentary, or mobile species respectively (Sheehan et al. 2010). A key 'indicator' species of recovery is the Pink Sea Fan as this is conspicuous and sensitive to disturbance and characterised by very slow recovery (Jackson et al. 2008).

A key aspect of this longitudinal work has been sub-contracting local fishers to provide maritime skills, local expertise and vessels to undertake the survey work. For example, a key early partner was John Walker, an experienced local skipper who supported the research project by building the sled used in the baited video, helped develop suitable methods to deploy and retrieve the field work kit, but particularly identified suitable survey sites due to his immense 'place-based' knowledge of Lyme Bay natural and cultural history. Operational surveys required equipment and skills associated with trawling such as handling the survey vessel and heavy survey gear, and selecting appropriate survey direction based on wind and tide. Consequently, a trawling vessel and associated professional crew was needed, a role undertaken by Skipper Kieran Perree and colleagues on the vessel 'Miss Pattie' (formerly skippered by John Walker). This chapter concerns the operational survey activity of a group of marine biologists from the UoP working collaboratively with these commercial fishers prior to, during the Summer survey season of 2016, and reflections on the associated 'social learning' involved. A particular survey was accompanied by the lead author on 13th July, 2016 as a participant-observer, and subsequently reflective writing and interviews were conducted by key members of the writing team.

Meet the 'Crew(s)' (the COPs)

Arguably, there are three pre-existing, but already overlapping, COPs relevant to this research:

- Marine Scientists – specifically, from the University of Plymouth within which a smaller sub-group has been created under the leadership of Dr. Emma Sheehan which has affectionately become known as 'Team Sheehan'. From this group a still smaller group of three researchers have been involved in the specific Lyme Bay survey research, Emma, Adam, and Amy (Fig. 11.3).

Fig. 11.3 UoP COP: (**a**) Emma Sheehan, (**b**) Adam Rees, (**c**) Amy Cartwright

Fig. 11.4 Fishers COP: (**a**) John Walker – former Skipper, (**b**) Kieran Perree – current Skipper

- Commercial Fishers – the Lyme Bay area has a number of fishers who, to a greater or lesser extent, have come together as a consequence of recent policy developments in the area under the auspices of the Lyme Bay Fisheries and Conservation Reserve. The two key 'on-board' fishers involved in the project are: (historically) John Walker (now retired); and (presently) Kieran Perree (Fig. 11.4).
- Marine Community Outreach – this work is largely facilitated by two organisations. The first is Blue Marine Foundation (BLUE) http://www.bluemarinefoundation.com/ which is a UK-based charity dedicated to creating marine reserves internationally, and establishing sustainable models of fishing. BLUE has been instrumental in setting up the Lyme Bay Fisheries and Conservation Reserve, and employs a local 'worker' Neville Copperthwaite. Crucially, Neville is a former commercial fisherman, and has therefore been employed specifically as a 'mediator' between the two COPs – marine scientists/conservationists and commercial fishers. In this capacity he facilitates community-group meetings involving the local fishing community and scientists; and also undertakes school-based outreach. The second is UoP. All three marine scientists are engaged in this work, but Amy has perhaps had greatest involvement (Fig. 11.5).

Fig. 11.5 Outreach COP: (**a**) Neville Copperthwaite, (**b**) Amy and Nick Higgs of UoP at Seaton Science Festival

Becoming a 'Devoted' Marine Citizen – Motivated by Love of, and Concern for, the Marine Environment

This section provides some reflections from the personal biographies prior to involvement with the project outlining key reasons for undertaking their career choices. A common thread across scientists' and fishers' reflections is the fact that they have chosen their career as a vocation. Certainly this is the case for all three Marine Scientists. Emma's passion started in early childhood:

> From as early as my parents can remember I wanted to be a marine biologist. By the time I was eight years old I was saying that I was going to study marine biology at the University of Plymouth. I never deviated from this plan and started my degree in September 1998 (Anonymous 2018).

Adam indicated that:

> Without really having a Eureka moment I somehow knew I would always be a marine scientist of some sort. From an early age a love of the sea has manifested itself through being a dedicated rock-pooler, to a kayaker, a surfer, a recreational angler and a keen photographer. … marine biology satisfied a need to be challenged intellectually while meeting my adventurous side too.

Similarly, Amy explains that:

> I had always wanted to work in environmental conservation but hadn't ever really found my calling. However working in and around the sea and witnessing the weird and wonderful creatures that inhabited it I realised I wanted to learn more about the science behind what I was seeing in the marine environment and work to protect it, so I embarked on a degree in Marine Biology & Coastal Ecology at Plymouth University and haven't looked back.

From the perspective of fishers, John noted that:

> For me, fishing wasn't and isn't about how much you made. The money comes afterwards, after all your hard work. It's more about the 'getting it right' that provided the real pleasure, especially starting from scratch like I did.

Neville describes himself as having:

> worked in the marine environment all my life in various guises, as a fisherman, diver, aqua culturist, sustainability marine manager and have been involved in setting up many projects, an artificial reef sanctuary, Britain's first commercial 1 lobster hatchery – for restocking – and a marine reserve being just a few examples. I consider myself a practical conservationist.

His statement reveals an interesting additional factor linking both the scientists and fishers discussed in this chapter, namely that most are also SCUBA-divers, affording them an additional 'window' (the diving mask) into the ordinarily hidden commons of the undersea environment, and a 'first-hand' experience of environmental degradation. As Amy recounts:

> I started university as a mature student after a number of years working as a SCUBA instructor in various locations around the world. I experienced first-hand the wonder of the marine environment and through teaching diving was able to share my passion and inspire others. I also witnessed the impact humans have on the fragile marine ecosystems that surrounded me; discarded fishing line covering coral reefs, destructive fishing practices such as dynamite fishing and untold amounts of plastic litter floating in the ocean.

The same was also true of John who developed a side-line alongside his commercial fishing in operating a SCUBA-diving operation:

> I must have first started diving at least 30 years ago. Back then I mainly used to take groups of divers out on Miss Pattie. … Devon Wildlife Trust, a local conservation body, went about undertaking diver surveys for a long time to look at this issue. I remember many good divers coming up and saying 'this is so sad', referring to the destruction being caused by damaging fishing practices.

And it was Neville's reflection on his unsustainable hand-diving for lobster that first instigated his awakening environmental awareness.

Another important aspect of the pre-project life-experiences has been the relevance of appropriate mentors. For both Adam and Amy, Emma has worked as their mentor into the world of marine science. From a fishers perspective, John recounted how, despite not coming from a fishing family, he was inducted by a friend of his father's:

> In his spare time Dad went angling with an old fella 'Alf' and his family and they would often take me. 'Uncle Alf' as I called him, the skipper, used to let me run a spinner and catch the bass which we would then cut up as bait. Alf used to hit me around the ear quite a bit 'watch what you're doing boy!', but I used to love it. Then many years later as he was getting older I started to take him out. He was like a second father to me.

However, it would seem that John learned largely 'on the job' through developing his own 'practical experiential knowledge base' discussed earlier:

> So I guess I started a fresh as my family were not generational fishermen, so all my knowledge and techniques were new, but I started to become successful. So then, naturally, I built a bigger boat which this time was a 24 footer which enabled me to 'fish through the lights' [fishing from day-to-night, or night-to-day]. Soon after building that I built my perfect boat from scratch, Miss Pattie.

Kieran also had an early-childhood through to adulthood mentor:

> I first got involved with fishing through my stepdad. When I was younger he bought me and my brother some pots so we went potting at age 10. I finished my engineering studies and then my stepdad taught me how to dive, so we both went scallop diving form when I was around 19 years old and I carried on from there.

Subsequently, Kieran, who was actually not from the Lyme Bay region but rather from the Channel Islands, met and developed a good working relationship with John whilst making long fishing expeditions to Lyme Bay; and the older, more experienced Skipper mentored him to the point where Kieran eventually took over captaincy of Miss Pattie and, consequently, also inherited a role in the Lyme Bay Research initiative.

Sense of Place, Place Attachment and Stewardship

An important dimension is the development of strong Place Attachment and a sense of Stewardship with *this* locality, namely Lyme Bay apparent amongst scientists and fishers alike. For example, John, as a long-standing resident in the area with a close engagement with the natural environment through his commercial fishing says:

> I have always been 100% environmental, especially after I've seen first-hand what we have in this area or Lyme Bay. It's my area, I've lived and worked here for a large proportion of my life and it has supported me. This interaction only enforces my interpretations as I now work to help support it. It's not just me that's fighting though. It's seeing others fighting to protect these areas that also continues to encourage me.

Similarly, a recent promotional article powerfully recounts how: "Through changing seasons, Dr. Emma Sheehan has come to develop a special regard for the 200 sq. km site, whose mud stone reefs shelter important fish and crustacean species. Whether surveying the sea state from the deck of the boat, or peering through the portal of video footage revealing the underwater life below, Emma now has an innate sense of the rhythms of the marine reserve." (Anonymous 2018). In the same piece she goes on to express this deep place attachment in her own words:

> I love the fact that I have a good mental map of the sea bed, and it's amazing to see how the different parts of the bay change each year … I don't think I was aware that I had such a personal connection to the place until we went to record the impact of the storms on the seabed in 2013/14. It was extreme, and I found myself getting quite emotional as areas of the seabed that had started to flourish had been scoured and covered by the sand washed in by the storms. (cited in Anonymous 2018).

Amy's relatively recent yet concentrated residencies in the area to undertake extensive scientific fieldwork have similarly engendered a strong sense of place:

> although living in the South West for the past six years (being from Norfolk originally) I'd never actually been to Lyme Bay there before working there this summer. Having worked and stayed in the area fairly intensively over our summer fieldwork seasons though I know it is an area that will always be close to my heart and bring back fond memories of the research I participated in there and the local people I met and worked alongside.

Then there are the reflections on personal practice – particularly powerful in terms of Neville's reflections on his personal unsustainable fishing practices in a nearby locality within Lyme Bay:

> I had been a commercial lobster diver working from Portland in Dorset. I had dived the Portland Bill area for 6 years and in the 7th, the lobsters stopped coming; I had fished them out. As I was the only person fishing in that specific area at the time I knew this overfishing was solely down to me. So I decided to do something to rectify the situation and started Britain's first commercial lobster hatchery and restocked with baby lobsters.

Shifting Perceptions of the 'Other' (COP), and Learning from Each Other

As noted in the introduction, there has often been a suspicion of the scientific community expressed by fishers. This was also in evidence in the Lyme Bay Case Study, perhaps expressed most eloquently by Neville:

> Since the beginning of modern marine conservationism in the 1950's the focus has been on the protection of marine flora and fauna. Humans were not considered as part of the marine ecosystem and from that base hypothesis it was a short and easy step to the demonization of fishermen as pillagers of the sea. … A mainly disparate bunch, fishermen were easy targets for well organised, well-educated conservationists. … People became entrenched and the 'us and them' culture flourished.

However, he recognises, and indeed is part of, a more positive trend in which marine biodiversity *and* human needs are more balanced:

> in recent years I have noticed the seeds of change. Food security has moved toward the top of many countries agenda and rather than pure marine conservation, more people are thinking in terms of sustainable marine management and are recognising the importance of the Blue Economy.

John came to this project with a long standing working relationship with scientists and recognises the potential benefit to his own marine ecological knowledge:

> I had been working with scientists long before Plymouth and I started together, so it was bread and butter and I knew what to expect. I always have said variety is the spice of life, so I was more than happy to be involved from the science side. While fishing you learn a lot about the ground and about the way fish behave, but you can never learn it all. I think any new knowledge can only be a good thing.

It was through John's relationship that Kieran became involved. After initial misgivings, he was won over:

> When I first took the job on I was worried and nervous. I thought they (Plymouth) were going to turn up on the boat dressed in suits and with briefcases. But after you arrived I realised you were just normal people from different walks of life. It was scary at first as I was initially testing new waters, I've always just done fishing so nothing like this before, but I'm glad we went down this route … I have always been interested in the science of scallops anyway, so it's an amazing job to be a part of.

As for the scientists, increasingly there is a policy expectation as discussed in the introduction, but also moral imperative, to involve fishers in their work, as Emma explains:

> I make a moral choice to use local fishermen wherever I can, because I want to involve those people who are so often impacted by decisions taken at government level. And when we send down our cameras they are often amazed by what they see – one fisherman in Hayle, Cornwall, thought there were crabs everywhere, avoiding his pots, and was surprised to discover it wasn't the case (cited in Anonymous 2018).

Adam also discussed the recurrent theme of the benefit to fishers of involvement over and above financial reward, including having a different view of the seabed:

> some realise the benefit of getting involved with universities and research institutions – financially and intellectually. Key example is Kieran – gets paid for the work, he knows it could be regular, and he gets to see images of the seabed…which as a scallop diver is cheaper and easier than strapping on a twinset.

Indeed, both parties recognise the reciprocal nature of the learning experience. A particular challenge was presented by the data-capture equipment (see Fig. 11.6). Emma shows how the science team relied on the practical-experiential knowledge-based expertise of fishers, both: in terms of handling the kit safely (demanding skill in safe deployment of heavy kit in choppy conditions); and the best places to sample (demanding a sound understanding of the marine ecology of the locality):

> It wasn't until I started to develop a new flying, underwater video system in Lyme Bay when I was not only relying on fishermen to teach me about their practice, but also to help me learn how use my equipment. It turned out that my new kit was similar to deploying towed fishing gear and so the fisherman I was using at the time was an expert and was able to show me exactly how to use the kit to record the best video footage. This involved learning about the tides, the wind, the seabed type. The fisherman even drew a standard operating procedures for me to take to my next study location in Guernsey. The system has now been

Fig. 11.6 The video array

deployed off fishermen's boats all over the UK to survey marine protected areas. Without the help from the fishermen, these methods would never have been as successful.

John also discussed this challenge, the necessity of working collaboratively, and the sense of fulfilment in problem solving:

> And then we move to the towing array, the camera system we developed to be towed behind Miss Pattie in order to look at the seabed. Again, 'it was fun to get it right! Haha'. The first day of testing it was 'like a bloody cockspring'. This bloody power cable kept twisting up, the camera was twisting around and we couldn't see the seabed! Myself, Emma and others spent ages getting it right on the first day. To get the height and weight right, to get the boat speed right and to work out the tidal conditions etc. But after a while it started to work, and we got there in the end. … I love making something that no one else has made and knowing it works well, its problem solving.

Kieran echoes this, and also points out other aspects of seamanship which the fishers are able to contribute:

> For example when we are working with equipment there are often problems but we come up with ideas together and create a solution that works. Because I work in and around the boat environment daily, I know how things work a little better so I can inform you guys with what I think would be best.

Emma indicates how the scientists depended on fishers' local marine knowledge, and were also obliged to learn good habits of 'seamanship':

> I also know where to go sample on certain day with Kieran, with video sampling it's important to get good video and this can depend on prevailing wind conditions, sea conditions and tidal state. So I go to Kieran at the start of each day and we make a decision together. He also helps us with ID as being a scallop diver he is accustomed to some of the same environments we look at. …
>
> I learnt how to stay clean and tidy. From the word go. Everyone on that boat was taught skills and about working on boats efficiently and safely. It's all very well just learning the skills but it's through being on that boat that you learn when apply yourself properly to avoid accidents and issues.

The fishers also learned a great deal more about the marine ecology of the area than their normal work-practices allowed, which had both intellectual and potentially financial implications. For example, John "learnt a lot more about salinity levels, water temperatures, water thermals, ocean currents and all those things that I knew little about."; whilst Kieran indicated that "my knowledge of the seabed and mapping of habitats has evolved far quicker than if I was doing it on my own accord through diving. Predominantly I have been focused on scalloping ground as thats my fishery but I'm always noting other things down too, such as suitable whelking areas for future reference".

Through these developing relationships it is true to say that there has now developed a strong sense of camaraderie, mutual respect and friendship amongst the scientist-fisher COP. Emma, in particular, describes these 'social bonding processes':

> By working together this changed quickly, and I think that we found mutual respect for each other. I think that I was quick to find this respect for John and it took a few years for me to prove myself to him. Nine years later, we trust each other implicitly. We confide in each

other and seek advice from each other. I ask John how to design and make new sampling equipment, and he asks me for advice on e.g. how to use IT, and for issues related to science and the marine protected area. When John retired, the new skipper Kieran was also keep to be involved with the university research. John taught him how to deploy the kit and gave us good character references. Kieran still had pre-conceptions about what marine scientists were like. But within a few weeks the trust between the new crew and new scientific team grew and new friendships formed. Three years on, I am now good friends with the skipper. Again, we confide in each other, and ask each other advice about working in the marine environment.

These positive 'bonded', more-than-working relations are often expressed through humour and shared 'banter'. As Amy says:

There's always a good atmosphere on the boat and although working hard the banter and jokes ensue!

Similarly, Kieran describes how …

I started to feel comfortable. We spend a lot of time on the boat together so naturally we end up having varied and different conversations, its' great! You guys are cool, interesting to talk to and fun to work with – 'you know I love you'

Learning to 'Border Cross' and 'Bridge' Between Scientific and Commercial Fishing and Wider Communities

These emerging strongly bonded relationships are the outcome of the social learning processes alluded to in the introduction which can be presented as in Fig. 11.7. Emma, Adam and Amy, along with their wider community of UoP colleagues, have

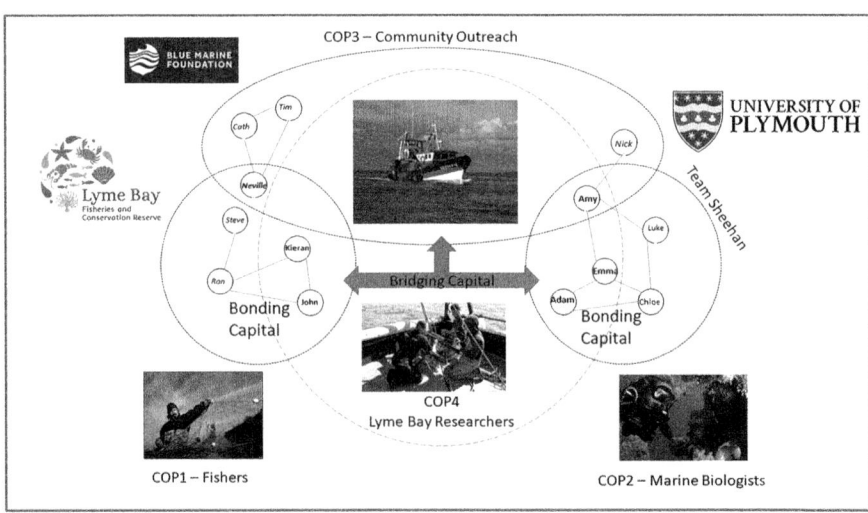

Fig. 11.7 Schematic of bonding and bridging capital within Lyme Bay Project

undergone a range of 'bonding' processes giving rise to a shared social capital relevant to marine science, thereby inducting them to a greater or lesser extent as 'marine scientists'. The community of fishers have, similarly, undertaken their own individual learning trajectories and associated social 'bonding' processes to arrive at a shared social capital which is quite different from that of the marine scientists. Finally, there are similar processes giving rise to an emerging COP of 'community outreach' workers drawn from both the fishing community (represented by Neville); others involved in outreach activities related to the local activities of the Blue Marine Foundation; and the marine science community (represented particularly by Amy who has worked alongside other UoP colleagues in specific Community Outreach activities such as the Seaton Science Festival).

Of key importance, however, are the 'bridging' processes across the diverse COPs which have been instrumental in 'creating' a new, COP of the 'Lyme Bay Researchers'. Keen and Mahanty stress the importance of communication: "If communication is a process of 'creating meaning' … then communication has a central role to play in learning processes that create knowledge across social divides" (Keen and Mahanty 2005, p. 114). This is in evidence in the clever ways that the fishers developed ways of communicating with the Marine Biologists. For example, in explaining some complex navigational boatwork, John resorted to a technical, yet humourous, diagram (Fig. 11.8). This represents one of the "new tools and symbols" (Keen and Mahanty 2005, p. 117) identified to facilitate co-learning.

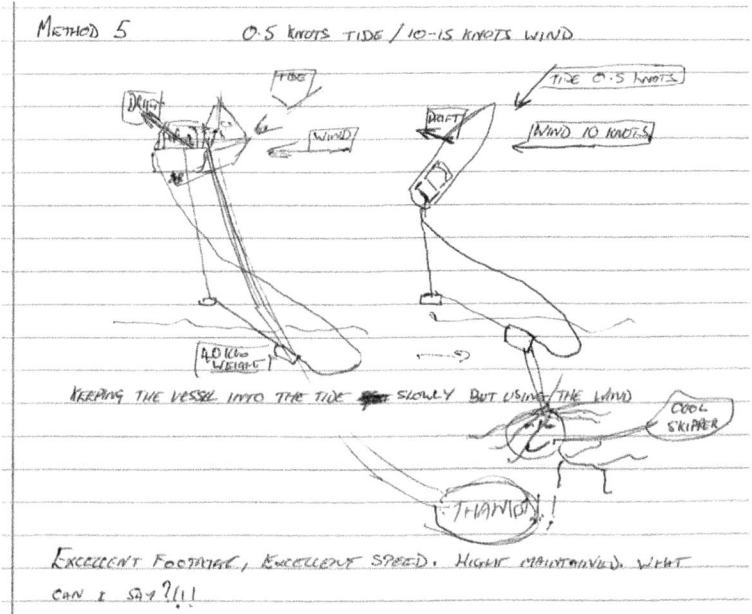

Fig. 11.8 Bridging communication device – fisher to scientist diagram

Casting the Net Wider – Community Outreach and Intergenerational Learning

The principle purposes of the Lyme Bay research is to provide evidence to inform marine management policy at governmental level. However, an important ancillary is community outreach which itself introduces a strong aspect of social learning. Here Neville occupies a key role as outreach office for the Blue Marine Foundation.

> I have been working with the Blue Marine Foundation who brokered the formation of the Lyme Bay Fisheries and Conservation Reserve on the South Coast of England. … The project has been a runaway success and the UK Government is considering this model of marine management as the way to manage all marine protected areas. The key to its success has been to include fishermen in the day to day management of the area and to include them in management decisions. The management system is called marine ecosystem based management. This simply means that we humans are taken into account and considered as part of the marine ecosystem.

A key part of his work is to facilitate communication and collaboration amongst key stakeholders. He achieves this partly through convening local stakeholder meetings with representatives from the marine science research community and fishers (see Fig. 11.9).

This kind of stakeholder engagement is important for getting fishers 'on board' with sustainable management practices. Thus, Adam recounts how:

> other fishermen have taken the opportunity to get off the boat and attend public engagement events in partnership with Blue in order to help promote the project etc. You can see them at a stand here at the local 'Dorset Seafood Festival', here they are promoting the idea of sustainable fishing, traceability of catch in order to provide the community with a better

Fig. 11.9 Neville facilitating a 'Stakeholders' meeting

product as well as getting better financial reward for catches. Something that Kieran understands, by adhering to a voluntary code of conduct introduced by Blue and getting his product – in his case hand dived scallops – recognised by a brand that represents sustainable practice and a better product at the end of it he can charge more money for it.

But the research is seeking to reach beyond the immediate commercial fishing 'stakeholders' into the wider community. Public Engagement and Outreach activities include communicating the principles and work of the Lyme Bay Marine Reserve and associated research through, for example:

1. the 'Seaton Jurassic' environmental education centre which has a dedicated outdoor play display which attempts to convey key themes in an engaging, fun manner (see Fig. 11.10).
2. Stands hosted by the Plymouth Marine Institute at the annual 'Seaton Science Festival' which utilise fun, hands-on activities to explain the research work and marine ecology of the area (see Fig. 11.11);
   and
3. 'Blue Marine Foundations' Schools Outreach programme delivered by Neville (see Fig. 11.12).

The social learning associated with these activities is beyond the scope of the current chapter and represents a desirable direction for future research. However, it is appropriate to present some of the work that has been generated by school children through the latter strategy which largely speaks for itself. For further details please refer to the website: http://www.lymebayreserve.co.uk/about/schools-outreach.php (Fig. 11.13).

Fig. 11.10 Seaton Jurassic's Lyme Bay Reserve interactive outdoor exhibit (author's daughter)

a)   b)

Fig. 11.11 (a) and (b): UoP's stand at Seaton Science Festival (author's wife and daughter)

Fig. 11.12 Blue Marine School's Outreach Program event (used with permission)

Fig. 11.13 (**a**) and (**b**): Exemplars of school childrens' work as outcomes of Community Outreach (used with permission)

Crossing Coastal Boundaries: Working Collaboratively Towards Sustainable Coastal Communities

Brown and Pitcher (Brown and Pitcher 2005) discuss a 'whole-of-community' pattern of knowledge represented as 'nested knowledges' comprising, each with "its own core body of content, ways of checking for truth, and well-defined knowledge boundaries" (Brown and Pitcher 2005, p. 128). These knowledges are: individual, local (common sense), specialized (e.g. that associated with being a marine scientist or, equally, a commercial fisher), strategic (the level at which policy goals can be established and worked towards), and 'holistic' which is the level at which a coherent vision can be formed. Arguably, the Lyme Bay Case Study reveals how these processes are enacted in a specific, Place-Based system related to the marine environment in such a way as to contribute to the overall sustainability of both the marine ecosystem and commercial fisheries. Hopefully, this Case Study reveals something of the complexity of the 'social ecology of learning' in this maritime location. And exemplifies the important bonding, bridging and 'border-crossing' learning processes at work as disparate, and sometimes antagonistic, Communities of Practice come together to work towards the social and ecological sustainability of a particular 'place'; and in so doing, develop ever more sophisticated individual and collective 'place-based' understandings and attachments.

References

Acott, T. G., & Urquhart, J. (2014). Sense of place and socio-cultural values in fishing communities along the English Channel. In J. Urquhart, T. G. Acott, D. Symes, & M. Zhao (Eds.), *Social issues in sustainable fisheries management* (pp. 258–277). Dordrecht: Springer.

Andrew, J., & Robottom, I. (2005). Communities' self-determination: Whose interests count? In M. Keen, V. A. Brown, & R. Dyball (Eds.), *Social learning in environmental management: Towards a sustainable future*. London: Earthscan.

Anonymous. (2018). Lyme Bay and Dr Emma Sheehan. *Invenite, 1*(1).

Bäckstrand, K. (2011). Civic science for sustainability: Reframing the role of experts, policy-makers and citizens in environmental governance. In S. Harding (Ed.), *The postcolonial science and technology studies reader* (pp. 439–458). Durham: Duke University Press.

Blount, B. G. (2003). Perceptions of legitimacy in conflict between commercial fishermen and regulatory agencies: Some ethical concerns. In D. G. Dallmeyer (Ed.), *Values at sea: Ethics for the marine environment* (pp. 127–146). Athens: University of Georgia Press.

Bourdieu, P. (1986). The forms of capital. In J. Richardson (Ed.), *Handbook of theory and research for the sociology of education*. New York: Greenwood Press.

Brown, V. A., & Pitcher, J. (2005). Linking community and government: Islands and beaches. In M. Keen, V. A. Brown, & R. Dyball (Eds.), *Social learning in environmental management: Towards a sustainable future*. London: Earthscan.

Brown, V. A., Dyball, R., Keen, M., Lambert, J., & Mazur, N. (2005a). The reflective practitioner: Practising what we preach. In M. Keen, V. A. Brown, & R. Dyball (Eds.), *Social learning in environmental management: Towards a sustainable future*. London: Earthscan.

Brown, V. A., Keen, M., & Dyball, R. (2005b). Lessons from the past, learning for the future. In M. Keen, V. A. Brown, & R. Dyball (Eds.), *Social learning in environmental management: Towards a sustainable future*. London: Earthscan.

Budzich-Tabor, U. (2014). Area-based local development – A new opportunity for European fisheries areas. In J. Urquhart, T. G. Acott, D. Symes, & M. Zhao (Eds.), *Social issues in sustainable fisheries management* (pp. 183–197). Dordrecht: Springer.

Davidson, L., & Stebbins, R. A. (2011). *Serious leisure and nature: Sustainable consumption in the outdoors*. Basingstoke: Palgrave Macmillan.

Dubois, M., Hadjimichael, M., & Raakjær, J. (2013). The rise of the scientific fisherman: Mobilising knowledge and negotiating user rights in the Devon inshore brown crab fishery, UK. *Marine Policy, 65*, 48–55. https://doi.org/10.1016/j.marpol.2015.12.013.

Eames, R. (2005). Partnerships in civil society: Linking bridging and bonding social capital. In M. Keen, V. A. Brown, & R. Dyball (Eds.), *Social learning in environmental management: Towards a sustainable future*. London: Earthscan.

Garavito-Bermúdez, D., & Lundholm, C. (2017). Exploring interconnections between local ecological knowledge, professional identity and sense of place among Swedish fishers. *Environmental Education Research, 23*(5), 627–655. https://doi.org/10.1080/13504622.2016.1146662.

Gieryn, T. F. (1999). *Cultural boundaries of science: Credibility on the line*. London: The University of Chicago Press.

Giroux, H. A. (2005). *Border crossings: Cultural workers and the politics of education* (2nd ed.). Abingdon: Routledge.

Gruenewald, D. A. (2003). The best of both worlds: A critical pedagogy of place. *Educational Researcher, 32*(4), 3–12. https://doi.org/10.3102/0013189X032004003.

Hildreth, P., & Kimble, C. (2004). Preface. In P. Hildreth & C. Kimble (Eds.), *Knowledge networks: Innovation through communities of practice*. London: Idea Group Publishing.

House of Lords. (2000). *Science and society*. London: HMSO.

Howard, P., & Pinder, D. (2003). Cultural heritage and sustainability in the coastal zone: Experiences in south West England. *Journal of Cultural Heritage, 4*, 57–68. https://doi.org/10.1016/S1296-2074(03)00008-6.

Jackson, E. L., Langmead, O., Barnes, M., Tyler-Walters, H., & Hiscock, K. (2008). *Identification of indicator species to represent the full range of benthic life history strategies for Lyme Bay and the consideration of the wider application for monitoring of Marine Protected Areas.* Retrieved from Plymouth.

Keen, M., & Mahanty, S. (2005). Collaborative learning: Bridging scales and interests. In M. Keen, V. A. Brown, & R. Dyball (Eds.), *Social learning in environmental management: Towards a sustainable future.* London: Earthscan.

Keen, M., Brown, V. A., & Dyball, R. (2005a). Social learning: A new approach to environmental management. In M. Keen, V. A. Brown, & R. Dyball (Eds.), *Social learning in environmental management: Towards a sustainable future.* London: Earthscan.

Keen, M., Brown, V. A., & Dyball, R. (Eds.). (2005b). *Social learning in environmental management: Towards a sustainable future.* London: Earthscan.

Lave, J., & Wenger, E. (1991). *Situated learning: Legitimate peripheral participation.* Cambridge: Cambridge University Press.

Leach, M., & Scoones, I. (2005). Science and citizenship in a global age. In M. Leach, I. Scoones, & B. Wynne (Eds.), *Science and citizens: Globalization & the challenges of engagement* (pp. 15–38). London: Zed Books.

Mackinson, S., & Wilson, D. C. K. (2014). Building bridges among scientists and fishermen with participatory action research. In J. Urquhart, T. G. Acott, D. Symes, & M. Zhao (Eds.), *Social issues in sustainable fisheries management* (pp. 121–139). Dordrecht: Springer.

Martindale, T. (2014). Heritage, skills and livelihood: Reconstruction and regeneration in a Cornish fishing port. In J. Urquhart, T. G. Acott, D. Symes, & M. Zhao (Eds.), *Social issues in sustainable fisheries management* (pp. 279–299). Dordrecht: Springer.

McCormick, R., Fox, A., Carmichael, P., & Procter, R. (2011). *Researching and understanding educational networks.* London: Routledge.

Morgan, A. (2012). Inclusive place-based education for just sustainability. *International Journal of Inclusive Education, 16*(5–6), 627–642. https://doi.org/10.1080/13603116.2012.655499.

NOAA (Ed.). (2013). *Ocean literacy: The essential principles and fundamental concepts of ocean sciences for learners of all ages Version 2 (NOAA Ed)* (2nd ed.). Silver Spring: NOAA.

Poe, M. R., Norman, K. C., & Levin, P. (2014). Cultural dimensions of socioecological systems: Key connections and guiding principles for conservation in coastal environments. *Conservation Letters: A Journal of the Society for Conservation Biology, 7*(3), 166–175. https://doi.org/10.1111/conl.12068.

Putnam, R. D. (2000). *Bowling alone: The collapse and revival of American community.* New York: Simon & Schuster.

Riesch, H. (2010). Theorizing boundary work as representation and identity. *Journal for the Theory of Social Behaviour, 40*(4), 452–473. https://doi.org/10.1111/j.1468-5914.2010.00441.x.

Sheehan, E. V., Stevens, T. F., & Attrill, M. J. (2010). A quantitative, non-destructive methodology for habitat characterisation and benthic monitoring at offshore renewable energy developments. *PLoS One, 5*(12), e14461. https://doi.org/10.1371/journal.pone.0014461.

Sheehan, E. V., Stevens, T. F., Gall, S. C., Cousens, S. L., & Attrill, M. J. (2013). Recovery of a temperate reef assemblage in a marine protected area following the exclusion of towed demersal fishing. *PLoS One, 8*(12), e83883. https://doi.org/10.1371/journal.pone.0083883.

Sheehan, E. V., Cousens, S. L., Gall, S. C., Bridger, D. R., Cocks, S., & Attrill, M. J. (2016). *Lyme Bay – A case study: Response of the benthos to the zoned exclusion of towed demersal fishing gear in Lyme Bay; 6 years after the closure* (Natural England Commissioned Reports, Number 219). http://publications.naturalengland.org.uk/publication/6330334199676928.

Snively, G., & Corsiglia, J. (2000). Discovering indigenous science: Implications for science education. *Science Education, 85*(6), 6–34. https://doi.org/10.1002/1098-237X(200101)85:1<6::AID-SCE3>3.0.CO;2-R.

Stebbins, R. A. (2009). Serious leisure and work. *Sociology Compass, 3*(5), 764–774. https://doi.org/10.1111/j.1751-9020.2009.00233.x.

Urquhart, J., & Acott, T. G. (2012). Constructing 'the Stade': Fishers' and non-fishers' identity and place attachment in Hastins, south-east England. *Marine Policy, 37*, 45–54. https://doi.org/10.1016/j.marpol.2012.04.004.

Urquhart, J., Acott, T. G., Symes, D., & Zhao, M. (2014a). Introduction: Social issues in sustainable fisheries management. In J. Urquhart, T. G. Acott, D. Symes, & M. Zhao (Eds.), *Social issues in sustainable fisheries management* (pp. 1–20). Dordrecht: Springer.

Urquhart, J., Acott, T. G., Symes, D., & Zhao, M. (Eds.). (2014b). *Social issues in sustainable fisheries management*. Dordrecht: Springer.

Wals, A. E. J., & van der Leij, T. (2007). Introduction. In A. E. J. Wals (Ed.), *Social learning towards a sustainable world*. Wageningen: Wageningen Academic Publishers.

Worster, A. M., & Abrams, E. (2007). Sense of place among New England commercial fishermen and organic farmers: Implications for socially constructed environmental education. *Environmental Education Research, 11*(5), 525–535. https://doi.org/10.1080/13504620500169676.

Alun Morgan is currently Lecturer in Education at the University of Plymouth where he leads courses on Environmental and Sustainability Education, Global Education and Outdoor Learning. He has worked in a variety of contexts over thirty years including as a schoolteacher, teacher advisor, lecturer and researcher in a number of Higher Education Institutions. He works across educational phases (primary through to Higher Education) and formal-informal learning sectors, and promotes intergenerational, lifelong and community-based learning. He has a long-standing research interest in Place and Landscape as integrative concepts for learning; and has a particular interest in Ocean Literacy and Marine Citizen Science.

Emma Sheehan is a Senior Research Fellow at the University of Plymouth. She leads a research team that studies human impacts (marine protected areas, fisheries/offshore aquaculture, marine renewable energy and dredging) on marine ecosystems to inform environmental policy and management.

Adam Rees is a marine ecologist undertaking research focussing on the impacts of various anthropogenic activities on protected marine habitats. His primary focus is on fisheries, having been involved with the Lyme Bay reef recovery monitoring since 2010. He has also contributed to research on the impacts of marine renewable installations and marine litter. He recently completed his PhD which assessed the impacts of commercial potting on reef habitats and the associated commercially important fauna within the Lyme Bay MPA. Currently, Adam is working with the Blue Marine Foundation coordinating research projects across multiple study sites throughout the UK, with the aim of providing conservation benefits and improving the sustainability of small-scale fisheries.

Amy Cartwright has been a research assistant on Dr. Emma Sheehan's team since 2015, supporting marine conservation projects ranging from marine renewables, sustainable fishing and marine protected areas. She graduated from the University of Plymouth with a degree in Marine Biology and Coastal Ecology (BSc Hons) in 2015 and after working in environmental education for a leading conservation charity directly after graduating, came to take up the research post at University of Plymouth later that year. Currently Amy's main focus is working on a project to assess how multiple wave energy converters interact with the marine environment, particularly determining any positive or negative impacts on seabed ecology and how these can be enhanced or mitigated.

Watch the 'Our Blue Planet' video about the project (available at https://www.facebook.com/bbcearth/videos/507954493003598/) and produce a concept map outlining a) the range of stakeholders involved in this project and b) the opportunities but also challenges of this way of working collaboratively across stakeholders.

Further audiovisual stimulus material related to the project can be found at: https://sheehanresearchgroup.com/return/, https://www.youtube.com/watch?v=Aefa4kAkVxU

Chapter 12
A Suspension of Reason in the Face of Food Harvest: Finding Meaning in the Act of Conforming

Michael Mueller

Ants are creatures of little strength, yet they store their food in the summer (Proverbs 30:25)

I was not born a farmer, fisher, hunter, or gatherer. I never aspired to be a harvester of food.

Thinking back to my childhood, I remember the bike that took me down to a nearby lake to fish for rainbow trout. My dad's old salmon pole swung side-to-side in my hand like a giant CB antenna as I weaved along the sidewalk. On the pier, older Native men who seemed to have retired there taught me how to use a double leader line with two trout hooks baited with salmon eggs. One day, one of them told me to spit on the hook.

"It will be good luck!!!" So I spit.

At first I had one but then two feisty rainbows on the same line. I still remember my heart pounding like it was going to exit through my chest and the old men laughing with me. Yes, two fish on my line and that's something a kid never forgets—for a lifetime.

Fishing was a nice distraction from a family that didn't have much money. Similar to most kids that grow up without much, my parents spent the majority of their time working so I fished on my own. I loved getting up early, riding down to the lake, and fishing into the quiet morning when the water was glassy. Sometimes I even skipped school to get there before my parents knew I had left. Trout tend to feed in the early morning hours. I casted until I dropped the bait in just the right spot near the lily pads. Trout fishing was a nice distraction, but I did a lot of things outside in addition to fishing. I loved to catch crawdads in a nearby stream. I collected different species of ants to watch them fight in a jar or dig tunnels through the dirt. I hunted butterflies (more on that in a moment). All I ever wanted to do was play outside. Playing outside wasn't something you needed money for.

M. Mueller (✉)
University of Alaska Anchorage, Anchorage, AK, USA

© Springer Nature Switzerland AG 2020
J. B. Pontius et al. (eds.), *Place-based Learning for the Plate*, Environmental Discourses in Science Education 6,
https://doi.org/10.1007/978-3-030-42814-3_12

I had little, if any, exposure to hunting. Hunting for game was never considered part of the food equation for my family. My dad never owned a gun. Not a single gun. Not even after my mother was held at gun point by a man who had just robbed a nearby store. I was too little to remember crying in the backseat. My parents didn't believe in guns. They said guns killed too many. Part of the issue was that guns were too expensive for my dad. So we didn't own any.

Although we didn't hunt, we always took our chances with the overgrown blackberry bushes hiding the chain link fences along many of the streets in Seattle. Blackberries made my lips and tongue purple. I loved to eat them while we picked small buckets full. My mom sent me into tight spaces to get berries while holding up the large vines so I could crawl underneath. She held them up, but I almost always got snagged by those goliath sized thorns on low hanging vines. My mother made a lot of amazing things out of the blackberries including blackberry jam and berry pies. Berries were the extent of my limited experience with gathering harvested food.

This chapter is an exploration into the story of finding the natural limits of knowledge and boundaries of the possible in harvested foods. This is a story of despair and hopefulness in the act of conforming and the suspension of rationality. It is a story about finding absurd ways to live and breaching the boundaries of what is known as food harvest that has afforded me the freedom to articulate my livelihood through local plants and animals. Many people do not live like me. They no longer hunt, fish, or gather. They shop at the grocery store. While recognizing this limitation for most, I do not always live like I'd like to.

And I continue to find more meaning in the act of conforming than not. For me, perhaps, striving for the harvest has been more important than harvesting itself. In the process of recognizing principles which guide my life I've identified a heuristic, and in this way, I'm better able to strive for what I value. I'll share the principles of this harvest heuristic with you at the end of this chapter as a way to cultivate a conversation of what might be pursued in the face of an oppressive harvest food paradigm.

Not for the Poor

Most kids will say their fathers taught them to hunt (MacDonald 2007).

My dad taught me to hunt for butterflies and other bugs. To kill them, we put a little nail polish remover on a cotton ball in a jar with the bug or butterfly and watch it go softly to sleep. Once the butterfly fell asleep we put pins through its wings and into a board. In this way, as the butterfly dried, the wings would become permanently fixed spread open widely so that we could admire our glorious trophy. It wasn't always this easy. In fact, I remember one time the butterfly woke back up and tore its wings from the pins that were piercing through. The wings ripped apart beyond repair even for a trophy and I remember feeling very bad inside. My dad put the little butterfly back in the jar with the fingernail remover for a while longer and we watched it go back to sleep forever.

Thinking about it now, these bugs were a lot like the fish that I caught. I rarely knew what to do with the dead bugs or the dead fish for that matter. I didn't know why I had to put the trout on a stringer but I felt compelled to bring them home to brag. I took as many as I liked without thinking about whether I was harming the population. When I brought fish home from the lake we usually admired them for a short time and then buried them in the yard. Eating a trout once for breakfast reinforced that I didn't like to eat fish as a kid. I wanted to catch them, see them up close, open them up and dig through their guts, blow up their fish bladder and admire their rainbow scales. There was no finer toy than the fish for a child who had very little.

By the way, much later as an adult, I learned that these trout were not endemic or wild. Rather they were hatchery-raised stocked triploid rainbow trout to hook eager anglers. In other words, they were raised primarily to hook (young) people on the value of trout or sports fishing. Hatchery raised fish may serve an important purpose in cultivating a love for fishing and nature. They were important in my life, but should not be confused as fishing for food (Mueller 2015).

I learned early in my life that eating seafood such as salmon, clams or crab was a luxury. I watched the fisherfolk throw salmon around Pike's Place Fish Market in downtown Seattle. I watched fisherfolk so many times and imagined being them. They were iconic Seattle. There were always tourists flocking to see fish flying through the air—cameras ablazing. I wonder though how many tourists actually knew what these fisherfolk did? I'd say, very few of them. So few people know how their harvest is intimately tied to the love fisherfolk have for the animal. I didn't really know the fisherfolk, but they knew me. They would catch me by surprise when I zero'd in on a halibut sitting on ice. As I crept up to take a closer look, one of the fisherfolk would yank a rope from behind the counter causing the fish to move and scare me to death! More than thirty years later, if you visit Pike's Place Fish Market today, you'll still see fisherfolk throwing fish around the market (and still scaring suckers). The people who buy them have money to claim a special status of being able to purchase seafood from the Pike's Place Market. These fish, clams and crab were foods that seldom made it to my dinner plate. They were not for the poor. The idea of seafood being something of a luxury has been with me for a long time.

Right about the time I started middle school I increasingly compared myself with friends who had money. I *became* poor. I realized it intensely. My worries became aspirations in light of peer pressure and thinking about how to escape from a family without much money—from being poor. My thoughts and actions became more pervasive as I grew older. I deemphasized my love for the outdoors for the pursuit of a good job, wealth and the ability to buy things without constraint. Similar to most Americans every aspect of my life was centered on the need to rise up from what I viewed as oppression in poverty. The way I now thought about food was escaping poor. In the act of conforming with the status quo that puts such a weighty measure on the high status for financial security, competition, and consumerism, I entered into a new heightened but obscure relationship with poverty that was imposed by my American education.

Let me explain. Many people are oppressed in ways that they do not realize. In deeply embedded ways the influences of our thinking – cultivated in the act of conforming – confine the impoverished (Freire 1970). Now consider the ways in which the deeply embedded cultural assumptions decipher and code the ways we live in relation to one another in North America. Consider the root metaphors of individualism, consumerism, economism and scientism to name a few. These concepts influence our everyday behavior in ways not often recognized (Mueller 2009). By the time I was in my mid-twenties, I had become so entrenched in the notion that eating and drinking should be done mindlessly that these things became as invisible as the air in my lungs. I did not think about the farmer who grew my food. I did not think about the places where my food would go if I threw it in the trash. I no longer even thought about the people with little food. Food itself became organized in my mind around events such as comfort food, friends, sports, movies, restaurants, and fast food lunches. I paid very little attention to the influence of advertisements that emphasized the processed food harvest. Eating food mindlessly and going to fancy restaurants hedonistically—Red Robin, McDonalds, Olive Garden, Applebees, Taco Bell, and Burger King—consumed my rationality. If there was any value for the food harvest early in my life, defined in whatever way you will, it would be found nowhere around me except in those I didn't know. In my mind, I wasn't harming people, farm animals, or the environment with my eating. And I would have told anyone who challenged this notion they were crazy. It wasn't just fisherfolk I didn't know. I didn't know the rancher, farmer, worker, beekeeper and a whole slew of others who provided the food that I so easily disregarded now. Conquering my poverty was a source of rationality where thinking about food was for the poor.

I wanted nothing to do with being poor.

Reclaiming Poor, Finding Despair

The seeds of love for the natural world were brought back into my consciousness when I began teaching elementary and high school. My students convinced me to create a hiking club and at one point it grew to more than 35 9–12 graders. While hiking Arizona's unique desert landscapes some of my students shared with me the ways in which their Mexican, Pima, Hopi, and Dine' heritages afforded insights into the desert environment. I became absolutely hooked on place-based knowledge.

Into my second year of teaching, I was called to active duty with the US Coast Guard. I was called in the middle of the year and it happened from my classroom. When I told my students that I would be leaving for active duty, they let me know how much they cared about my teaching. At that moment, it hit like a ton of bricks that I needed to be much more cognizant about what I was doing as a science teacher. I had to find a different way.

When you are in the Coast Guard you learn to surf because it is something you can do with your crew and there is always a beach with surf break nearby. So I bought an old $30 surfboard and patched up the holes and dents with an epoxy repair kit. Learning to surf was a lot more challenging than I thought. It took me more than two months just to stand up. Indeed, I felt like such a "kook" (a word for someone who poses as a surfer). As bad as I was at surfing, it took my mind off being deployed. It took my mind off teaching. It was this renewed learning that pulled me back towards my aspirations as a child who loved the land and sea. Towards a relationship that captivated my childhood, hiking with youth who saw the world differently and dancing on the ocean started to erode and dissolve the mindless way I viewed the food harvest.

When I returned to my home in the desert, I felt trapped in a scorched habitat where I could not escape the stifling metropolitan heat and urban sprawl of the city. I tried to grow tomatoes but they rapidly wilted. It didn't seem right to grow human food in the desert. The farmers who grew cotton, roses, and vegetables on the outskirts of Phoenix irrigated their land by way of an aqueduct that ran thousands of miles from the Colorado River. Grocery store food was also delivered by truck from thousands of miles away—the closest place being California.

As I returned to teach, I felt the need to teach in a way that would help my students become cognizant of the conventional harvest and the ways it taxed the land. But there was no food harvest curriculum and even environmental education was nowhere to be seen. When I mentioned food to my teacher colleagues, they pointed me to a teacher at another high school who wrote several grants to build a greenhouse and cultivate a school agriculture program. He raised hundreds of tomatoes with his students and sold plants and fruits to the community. It was right about then that I decided some of my teacher colleagues had retired in the classroom and that they also resigned to the constraining rationality of preparing youth for consumerism. Like these youths, I had once been at the mercy of the decontextualized curriculum of schools.

So I left teaching middle and high school shortly after returning from the Coast Guard to pursue a doctoral degree. I felt the rekindling of a flame that had simmered inside me for years. I recognized the need to foster this natural inquisitive nature about the world through curiosity with my children and youth in public schools. I started to believe that nature may even mediate the ways in which people eat food mindlessly like zombie walkers and the rationality implicit in the corporate farming paradigm. So I learned everything I could about the ways in which food is gathered including farming practices, animal husbandry, beekeeping, impacts for farm workers, and I started to grow food in a little garden (it died but I tried). More, I played outside with my son as much as possible.

What I had welling inside made matters worse.

Becoming again, I was stuck. We had just moved our family from Arizona to Tennessee. With very little income, the only way to get by was to partly live on student loans which meant eating the cheapest food available. This time however, I faced despair. It felt wrong. It felt ridiculous to buy processed and conventionally raised vegetables that I knew had been genetically modified, sprayed by herbicides and pesticides, and required immense watering. My wife and I talked many times about the food we were feeding our family. In deliberation, we started to visit a barely thriving farmers market downtown Knoxville with locally grown foods. The farmer market felt like home. We connected with people who visited the market to just hang out, listen to music, and talk with farmerfolk. It was increasingly painful to buy food from the grocery store as I learned more about harms and perils for farm workers and their animals.

Discovering the Politics of Food Harvest

Conventional food harvest has become dressed in a cloak of consumerism. The politics of consumerism serve to mask the conventional harvest for the rationality of food shopping and processations basically out of our view. Any tinkering with hobbies such as gardening, fishing, hunting, or gathering are eclipsed by the supermarket where sundry chemicals and unethical animal farming practices remain invisible. Foods are labeled genetically modified or genetically engineered, organic, healthy, free-range, or cage-free with reduced or recycled packaging. Stores ban plastic bags or charge extra. Store logos or cartons display happy cows, pigs, chickens, bees, and vibrant disease- free or bug-weed-spray-free crops. When it comes to food, Americans spend the smallest portion of their budget at the supermarket (Singer and Mason 2007) and yet food is the most fundamental part of life itself. Consumerism functions to mask the wild places where food might be harvested for free or without a lot of money. It is as if these places were extremely dangerous and out of bounds epistemologically—"You would have to be crazy to get your food that way!" Today's food harvest employs defenses such as changing advertisements, coupons, displays, and labels. Food changes oppress in a new way, unrecognized by most. The rational choice by consumers is to embrace the stores that cater to the latest trendy social issues of our society. And in thinking that stores are making strides towards these issues, the hopes are that (we) (the consumer) will continue to view harvesting one's food as unrealistic. Most people do not think about going to the grocery store as a form of oppression because there is prudence in the visible food harvest. The politics of food harvest produce ridiculous poverty in the face of affluence and rationality.

Gathering Food from the Land—Becoming Farmerfolk

Small farms are becoming increasingly invisible in North America as schools prepare children for urbanization and leaving the farm that has been the story of the last century. The perpetuating stereotype is that farming is a difficult business and there is not much income in gathering food. On one hand there is the despair of living with so little money, where small family farms run up against the shenanigans of corporate farming. But on the other, there is a sense of freedom. The limits of the possible can be explored while the land offers wisdom guiding nonconformity. While there is a mounting expanse on the organic market farming scene from youth especially, farming is still a very difficult dream for most people who chase after American consumerism.

We decided to move to a farm after living several years in Georgia where I had taken a job as a professor teaching teachers of science how to engage students in science, place-based education, and food justice. Neither my wife or I had any experience farming and we really only knew that we wanted our children to have more experiences running around on the cultivated land and developing a love for animals. The farm turned out to be a two- pronged project of learning about the local Appalachian ways of knowing around farming and animal husbandry for my job, and a way to grow our own healthy food.

Eventually the farm would become a place to grow food for 100 other families across Atlanta (Mueller 2014). This experience farming was characterized by two years of learning to farm organically, learning from local farmers, talking with farm store owners, visiting farms, reading farming materials and just putting my hands and the hands of my family in the soils around us.

During this time, I went through a hectic experience of tenure and promotion at the university that created intense wounds. The farm served as a refuge. The land provided a playground for my children and continued to renew my love for the biodiversity of nature's harmony. Outside of academia, we lived a farm inspired life rich with observations, experiments and hard work. We grew responsibly nurtured vegetables, herbs, fruit, and flowers for a downtown farmers' market. The farm imprinted as part of my fabric or what I call socioecological character (a way of being in relation). When we entered farmerfolk culture we did so as a family. Today I would say strongly that this integrated work- life-balance challenged academic career dreams and made it very difficult to navigate the tenure and promotion process. Despite that I had published almost thirty papers, many of my colleagues would not accept that I did not publish but a few of them in mainstream places or that I lived my research in a way that seemed absurd and ridiculous compared with their work. My work consisted of a newly envisioned ethnophilosophy, where we kept detailed garden and field plans, seed journals, personal journals, took thousands of digital photographs and created social media to illuminate the farmerfolk experience. While I battled tenure and promotion as we became farmers, the thought of continuing in higher education began to fade.

I'll explain further. Teacher educators do not make a lot of money as compared with their colleagues in other parts of the university, especially business faculty. Saving income during the year for the time in the summer when faculty are off contract is not easy. Often I would have to find other jobs. Becoming farmerfolk meant something entirely different. Although summertime is the peak time for selling organic produce at the farmers' market, we would need to embrace the thought of making sometimes as low as $300 dollars and as much as $1000 a week. The cost of maintaining equipment, fuel, seeds, rental property, utilities, and so forth, ate up most of the budget leaving us with very little income to pay for student loans and other debts and expenses we incurred before the farm. It was during this time that we began to realize how tethered and trapped we were in the consumerist culture of poverty.

We entered the farm life to become part of the counterculture of "back-to-the-landers". The farm helped me to engage my students at the university in thinking about their relationship with harvested foods: that their food choices came with consequences and that we have a responsibility to select food wisely.

Because our farm was a responsibly nurtured place, food could be gathered and eaten right off the stalk or vine, and we did not worry about fertilizers, pesticides or herbicides within our bodies when we harvested fruits, vegetables and flowers (by the way, there are a lot of edible flowers.) We came to realize that freshly harvested food did not have to be gathered within a particular region, or within a season, or even before the time it was ready. It was comprised of a *relationship with the farmer*. Our customers or people who visited the farm and farmers' market developed this relationship with us over some time. My students also realized when they visited the farm that there is more to the idea of fresh. It sharpened their decision- making in relation to food just like it forever changed my family's relationship with gathering.

The farm taught polysemic knowledge, embodied change and positionality, and environment condition. These themes will continue throughout my story. Before moving on to what these things are, I need to acknowledge that people struggle with food scarcity worldwide and that grocery store food may be all they have to survive. I'm not taking the high road here. I know that people make sacrifices every day to support their family. What I am exploring in this story is the idea that I have not been perfected or reformed in the absurd act of nonconformity. It was breaching the limits of rationality that was experienced through the farmerfolk lifestyle that better positioned me to engage with the food harvest in a way that would change my life.

Polysemic knowledge is cultivated in the reflection of thought and experience. On the farm, this reflection came through an embodied relational characteristic with place and by way of increasing awareness around seed saving, growing vegetables, health benefits of greens, and renewal of intergenerational knowledges and experiences, narratives, traditions, and plans which account for the future of a place. We recognized the shards of local history embedded in the soil (pottery, arrowheads, and tools). The soil had a story to tell if we just listened. Written and oral knowledge contributed too, such as what we read in the Farmers' Almanac, anecdotal accounts of companion planting, scientific reports, mathematical charts, Appalachian history, planting skills, and Native American knowledge. Working the farm field took time and appreciation, and positioned the value of particular knowledge. Playing as a family in the dirt, cooking and creating meals, and preserving vegetables and fruits fostered intense bonds. The physical geography, geology, climate, weather, sunlight, shade, runoff, dry and wet areas for planting, plant disease, and insects all had their say. Clothing and lotion choices, to protect the skin, bodily ailments and limitations, insect bites and stings, and medicinal treatments shifted the evolution of boundaries. Learning about polysemic knowledge occurred everywhere but most intensely at the farmer market where we interacted with people through the seasons. The market itself was constituted of food, cooking, exercise, music and the aspirations of those people who would never travel closer to the farm than the farmers' market in a downtown park. We shared photos and experiences with our patrons and they provided a glimpse of their world. People brought their children, dogs and other pets. One boy brought a different lizard, insect or other creature he had caught in the park. He reminded me of the boy I once knew. We interacted with youth, people who were pregnant, gay, mother, immigrant and so forth. We documented how they relied on their own life aspirations, experiences and science knowledge to navigate food choices. The farm market was a place for all walks of life, yes, even celebrities. (We once met Owen Wilson riding his bike through the park—he said we had nice tomatoes!)

We also learned embodied knowledge and change from the hardships faced on the farm. I tore the fingernail clear off my finger while pulling beats out of the ground. I have so many Fire Ant scars I'll never forget how the land breathed fire beneath our feet. There was always the question of "whether this is all worth it?" There were so many unexpected things that would happen on the farm. The truck would need repair, insects would take over a particular crop, or the weather would dry up or it would get too hot for flowers. We would plant something in the wrong place or think we needed to grow everything we ate. Eventually we learned to trade with other farmers who grew things we didn't have for vegetables, fruits or flowers that we had. The flowers we grew attracted different insects throughout the year. Being in the thick of the farmerfolk culture become as much the advocacy of a positionality and agency as it was a way to sell healthy food and get to know those people who participated in our farm.

Environmental condition is the work of monitoring the predicament of the environment.

We learned and used scientific knowledge about weather, insects, soil chemistry, seed saving, and preservation methods. We explored citizen science with the kids to think about change. Not only did the environmental condition contribute to broadening the boundaries of what we knew but interacting with people at the farm market offered perspectives on food mindfulness.

People often mentioned some nostalgic experience associated with the food that we had or how the flowers reminded them of some significant event from their past. Perhaps a mother kept flowers around the house and this particular variety was her favorite. People purchased flowers and food for cultural events, birthdays, anniversaries and other celebrations. They purchased food because they wanted to experiment with cooking, or live heathier for a future baby or for unexpected health problems. People learned about the medicinal characteristics of the plants or the nutritional properties and vitamin components of vegetables or herbs. They also bought produce that would attract their children to the regionality and flavor of Georgia.

There was never any waste on our farm. Food was eaten, fed to other animals, or composted. We learned to appreciate seasonality, regionality, insect impact, and food disease. The true value of every vegetable, fruit and flower was fully understood. The ethical treatment of plants, animals and people had become more visible than ever before in my family and life.

Most importantly we learned that people became more mindful of the food harvest when it grew into something special. *Hopefulness.* This hopefulness is a lot like one's love of nature. We tend to return to what we find special in our lives. It is a relationship of hopefulness. We had become so enamored with being farmerfolk that at one point we contemplated leaving higher education for the farm life altogether. Unfortunately, it would never have paid the bills. It wasn't in our cards. We had too much debt from the past and the pursuit of hedonistic values in a university job. Our hopes were fragmented. Some of my colleagues continued to make things miserable. With promotion finally earned we took a different job 4000 miles away at the University of Alaska. This time we had no idea how we would gather our food. We knew we had to try. Moving to Alaska intrigued me because of the ways that I imagined Natives and Alaskans living with the sea as their garden, but also gathering berries and harvesting game from the land. I had no idea how we would learn what we needed to survive off the land but I was encouraged by the aspiration and hope that came with thinking about how to reframe food in this place.

Gathering food from the sea—becoming fisherfolk The first year in Alaska was exciting but challenging food wise. When we left the farm, we sold just about everything we owned. We arrived in Alaska with boxes of memorabilia (1500lbs). We rented a house in a neighborhood where the houses were 3–4 ft. apart, a hundred on one street. Not what you think of when you think of Alaska. We slept on air mattresses for the first month and scavenged garage sales for furniture and other household items we needed to make a life. To become a resident of Alaska you have to wait one full year. Once you become a resident you can participate in subsistence fishing and hunting (carried out to put away food for a family). Until then we were stuck buying food that had been shipped by barge or plane thousands of miles to the grocery store.

Many people spend only a short time in Alaska. They come here for many reasons. Most of the population of Alaska lives in Anchorage, the largest city, where urban conveniences and grocery stores are aplenty. I learned there is a certain romance in living off the land, but most people do not see the value in doing it. Even in rural Alaska it is often easier to go to a store.

While there is some interest in an emerging Alaska farm movement which supplies some root vegetables, leafy greens, and potatoes, most produce comes from the lower 48. We found a Community Supported Agriculture (CSA) project out of Washington that shipped food up to Alaska but the shipping was very expensive. Eventually we could not keep up with the CSA and we returned to the grocery store where food is available anytime of the year on the cheap. The price of shipped food increases dramatically as it travels further away from Anchorage. For example, a gallon of milk can cost $3–4 dollars in Anchorage and $10 dollars in many villages.

While living off the land can reduce the cost of living in rural Alaska, it is not necessary in Anchorage and communities on the road system (Wasilla, Palmer, Fairbanks, Kenai City). The vast majority of Alaskans are dependent on the supermarket to meet their needs. However, a good number balance that equation with fishing, hunting and gathering. Some live 100% off the land. While I was waiting to become a resident, I fished the only way that I knew—by rod and reel. This was the first time in almost twenty years that I held a fishing rod. I purchased my rod, line, hooks, floats, stringer and hip boots from a garage sale. Determined, I put 35 salmon (including one "King" or Chinook Salmon) in the freezer catching them one at a time, with a limit of three fish daily for "Silver" or Coho Salmon. Once again I reclaimed that boy inside me. We realized that silvers didn't really freeze that well and tasted horrible after being frozen for more than a few months. Kings and silvers are best eaten fresh from the ocean and better if smoked. My wife and I were desperate to feed our family through harvested food and we ate those juicy nasty silvers with Cajun spices, dill, lemon and any other recipes we could garner from friends.

We also gathered blueberries high in the mountains above town of Eagle River. One thing about Alaskans is that they know about a lot of secret spots that need to stay secret because of the ways that nonresidents and tourists will treat them if they learn where to go. So it goes with blueberries. There are tourist spots and locals' spots. The size and quality of the berries differs dramatically. The first couple years we started with tourist spots because we didn't know how to scout for berries or where to even go. We could have gathered enough berries to last a few months, but they were gone in weeks. Wild berries are something of a commodity here in Alaska and there are lots of different blue, red, orange, and white berries that grow wild in different bioregions and elevations. Particularly important are knowing what berries to eat and what not. Some berries will kill you! And there are lots of tales that Alaskans will tell you of someone they know who died from eating the wrong berries.

My time as nonresident went fast and I focused on food justice with my new students who taught me a lot about Alaska. Some of them even took me fishing and discussed places to begin hunting. It felt good to be an Alaskan resident! To gain a glimmer of hope in the face of despair. Somewhere between the despair of conforming again with supermarket food and the hopefulness of striving for the freedom gained through fishing, hunting, and gathering for food, were the situated tensions associated with becoming. Becoming is something we can all strive for even if being happens on the boundaries of what we consider rational. Striving for and in the absurdity of becoming there is a renewed hopefulness. For me, that hopefulness eventually assured the guidance and freedom to mediate a very oppressive food paradigm. Hopefulness provided new understandings at the boundaries of what was considered knowledge. My story is a creative passion and love for the natural world that now fuels my pursuit of harvested foods.

Hopefulness was a pursuit that made poverty absurd.

Being Fisherfolk

As a family we take a yearly trip to Kenai North Shore Beach on the Cook Inlet to dipnet for Sockeye Salmon. My two youngest children have participated in the harvesting of salmon since they were three and four, and this is all they know (my daughter Summer is 7 and son Noah 9).

The water is very cold but here you can see Alaskan children do what they know and even swim in the frigid water for fun! The sun is closer to Alaska at this time than anywhere else on the planet. We pay very close attention to our exposure. Harvesting close to our limit of 65 "Reds", we camp on the beach for four to five days or until we've breached our physical limits. The sand is hot and the surf is cold. Standing in the water with a dipnet for hours at a time during the incoming and ebb tides, there is little attunement of the ways the chilling gray clouded water seeps into your bones. I know I'll be sitting by the fire later drinking a beer. That will eventually provide some clarity for the next time I don my cold soaked waders and wader shoes. For now, my mind is numb. I'm patiently waiting for the next salmon to swim into my net. The thrill of the net's vibration and salmon's wild fight will exhaust me. The glimmer of silvery scales will entice me.

"I've got one!!!"

"Go dad!!!" "You still got him?" my oldest son Riley yells.

As I'm dragging the net through the water I can feel the charging current moving back to the sea. I move as swiftly as I can. In the distance I see the shore and sand where my wife is waiting with my jumping young daughter and son running back and forth—"dad you got one?"

They are ready to work. I pull the fish carefully through the turbulent surf, where I've lost a few salmon. As I look back, I see the glistening silver shine blister and water spray. The fish is still in my net! Exhausted and breathing hard, I pull the rest of the net onto the sand. A blue-silver Sockeye salmon flops and wriggles in the net entangled by the fight. Other fisherfolk are pulling in their catch too and as I scan the shoreline there are a good number of fish caught. One salmon usually means more in the nets. Now up on the beach sand I run the length of the pole to the net or about 20 feet, to where I can club the salmon over the head with my wacker. A fish wacker is used to put the salmon to sleep. I know right where to slap the head of the salmon. It shivers in a sporadic wiggle that signals death. I pick up the fish by the gills. My wife takes the fish and pulls out the bright red gills snapping blood on my waders. Blood runs quickly down the fish's scales towards the tail. She hacks two cuts in the tail that indicates the fish has been harvested for our family (by law) and puts it in a bucket of seawater. I head back out.

As I look down the beach towards the water, the sunlight skips off the ripples of the waves and seagulls haphazardly glide by. I see Riley's dark silhouette pulling back on his net.

"I've got one too!!!"

Fishing for Food

Subsistence fishing is a meaningful cultural activity where families work together to take what they need for the year. There is a lot of comradery and spirited exchange as fisherfolk cheer each other on, especially for the young apprentices who are just beginning to help with the harvest. For many Alaskans, dipnetting signals a time of year and place. During the height of the summer the sun shines almost 22 hours. The midnight sun breaks only for a few hours during our sleep. We all look forward to being part of this harvest with our families. Many Alaskans describe dipnetting as the best week of the year. A full harvest will sustain most families about 2–3 times per week for the rest of the year. The work just begins with the harvest. We have to gut the fish, fillet, and pack them in ice. We throw the guts to the seagulls. They fly in droves above our heads. Cleaning fish usually happens at the lowest tide. The leftover carcasses are tossed to the gulls and what remains is washed out to sea for the crabs and other scavengers.

Sockeye meat is hearty and rich. It holds up well in the freezer. The real work begins when we get home and prepare the fish for the smoker or the freezer in vacuum seal bags. While processing the salmon, we might find little lamprey like organisms called sea lice and inside the meat we occasionally find long transparent tapeworms, which are common in ocean- caught Alaskan salmon. When cooked, smoked or frozen for more than six days, they are no longer a harm. During the year, my family will eat 4–5 filets of salmon a week for five people. The market value of these 65 fish ranges from $10,000 to $15,000 depending on what time of year they are being sold. My family saves more than $10–$20,000 yearly by harvesting Chinook, Sockeye, Coho and Halibut that we either dipnet for or catch by rod and reel (or sports fishing).

Hopefulness in the pursuit of fishing for food has cultivated a different rationality for my children who all have taken an interest in fishing. Mediating the conventional food paradigm by passing meaning in fishing for rainbow trout on to my kids, I'm conserving the intergenerational knowledge that was once afforded for me on the pier. We can't fish with two hooks on one line, but growing up fishing ensures a mindfulness and hopefulness in striving for the harvest that brings forth a healthy awakedness. My children are not always as excited as I am about eating the catch, but they understand intimately why we take the fish apart and what they taste like. We thank every fish and they understand the significance of a fish that gives its life for ours.

Being Hunterfolk

Hunting is not something you just fall into. Alaska is one of the more difficult states to get hunting tags because of the lottery system that is based on pure luck. You put in for tags and hope that something comes from it. On the other hand, there is a subsistence (Tier I tag) that provides at least one caribou per family per year. If you register for that tag though, you are limited to only one caribou per family and no other game. For many people who hunt for food, a subsistence tag is enough to offset some of their grocery store dependence. They take advantage of the subsistence hunt which keeps the Nelchina Caribou Herd at bay with 3500 animals. I always put in for other tags because if drawn, your chances of hunting successfully are increased quite a bit. If I don't draw tags, then I just use the general harvest tags and with hard work these can also provide successful hunts.

Getting to know the behaviors of large game animals takes a lot of time in the tundra and mountains. It is important to understand the behaviors of these animals because they can be very dangerous. A full grown moose can weigh as much as 1500lbs and a bear can charge over 35 mph. There is nothing more dangerous than a sow bear with cubs or a cow moose with calves. Finding game in Alaska is also difficult because the population density is less than most of the other places hunted outside of Alaska and there is a lot of land to hunt. This last October I registered for an "any bull" moose tag imagining that I would hike over the mountain summit and six miles into the backcountry to harvest a moose. After sighting three moose down in the valley on the other side of the mountain my hunting partner and I walked three miles through the snow and followed the tracks into the woods. The woods make things interesting because you no longer have the advantage of wide-open space to buffer the charge of a bear or moose. We ended up finding a cow moose and her calf. The third moose was probably a bull in the rut (when breeding occurs). Frustrated with the cow, it probably raced off further down a creek that ran along the ravine where we were hunting. I wasn't going to harvest the calf. Disappointed after such a long trek we ate lunch near the moose and watched them chew on alder buds and spruce tips. Spending that time taking in the hunt made it worth the trip.

Thankfully, my hunting partner and I harvested a spike fork moose a few months prior. He texted me one evening from a spot I had told him about.

"Moose down Mike!!!"

"I'll be right there" I texted back.

I was eating dinner with my family at the time, and it was getting dark. But I knew we were in this together. I hiked two miles with a pack, sled and wagon, and over 1700 feet elevation change in the darkness to find him. It was pitch black! There was very little my flashlight could do to infiltrate the darkness. I called out to him and finally found him about a quarter mile off the trail near a deep ravine. Had the moose taken one more step there would have been no way to recover him. We worked swiftly to get the moose into game bags. The first part of the journey would be incredibly tough. We had to pack the moose out, 350lbs each. We hiked 700lbs of meat out together over 4 h. Covered in blood and shaking from the cold, I was thankful not to come across a bear. It was the most exhausting thing I'd ever do. One young moose is enough for two families to share. Every time I eat that moose I think of harvest.

Hunting for food Harvesting a large game animal is much more about a relationship with the land than it is a hunt for something to kill. Before we harvested that spike fork we spent more than a week hunting near Fairbanks, which is about 6 h driving distance from where I live. The general harvest tag required that the moose be a bull with an antler span of more than 50 inches or a variable number of brow tines depending on the region. Glassing for hours and hours, we saw very few animals.

When we did find a moose it was never legal. They would peer intensely at us with their defiant brown eyes knowing that we could not harvest them for food. When you are hunting there is an itch that needs to be scratched, the intense desire to touch a beast. Although you will tell yourself at the time that the hunt is worth the experience even if there is no animal to harvest, it is almost as if you are deceiving yourself to conceal the disappointment. When I think back however, it has always been worth the time invested and lessons earned on the tundra or mountain. The moment a large animal finally presents itself there is a paralyzing shock to the body. Blood pressure rises and eyesight becomes keen, and there is predatory meeting of emotion and epistemology. "Is this animal legal?" "What shooting position should I take?" "Is there an ethical shot?" "Why take this animal?" "Do I really need this animal?" "Can I physically get the meat out of this place?" Waiting for the moment when an animal is broadside or lifts its leg just enough to preserve most of the meat can feel like hours wrapped up in a minute. Although I never remember feeling the pulse of the gun or the deafening blast, I hang on to every shot as a significant event and replay the last steps of the animal for several years after the harvest. While I'm not married to the idea of capturing death in my hands for a photograph, I have taken that picture before in the childhood spirit of a trophy butterfly that epitomizes my first hunt. I need to touch the beast, touch the animal's fur and touch its body to reach the climax of a hunt. Sometimes I cry. The encompassing harvest silences the rush of emotions and calms my nerves. Mindfully engaged, the first cut with my knife is always the most difficult to take. I feel as vulnerable to the absurdity of butchering meat for my family as the blade gliding under skin. Suspending reason leaves deep cuts that turn into scars. The harvest allows them to fade.

Epilogue

The hunt is never entirely about the harvest of food, just as it is with fishing and gathering. I've started to raise bees and thought about building a greenhouse at 2000 feet elevation where I live in the mountains. We have raised beds, but there are just some things we can't do such as composting food (although I've tried a worm bin with some success), and there are some things that would be considered unethical like using fish carcasses for fertilizer. There are too many black bears and coyotes that have names that live around our home. At the same time there are lots of things that I can strive for that will take finding meaning in the act of conforming.

This past year we fished for King, Red, and Silver salmon, halibut, and rockfish. We hunted for moose, caribou, and traded for bear. We gathered over 100lbs of wild mountain blueberries, cranberries, and fireweed. We grew raspberries, radishes, carrots, lettuce, chard, broccoli, cauliflower, snap peas, and tomatoes. We made fireweed jelly, blueberry jam, and raspberry jam. We made blackberry jam from blackberries we picked and canned 150 pickles (from a trip that we took to Washington State). Finally, we made a Melbec and blueberry wine.

We tried beekeeping for the first time with some success. Last year, we left the honey on the bees for the winter. Hopefully there will be a harvest this summer. We are planning to raise chickens this summer for meat and eggs. We do make bread, but buy the flour, yeast and salt.

We still rely on the grocery store for many of our vegetables, fruits, milk products, pasta, sauces, sugar, cat food, coffee, beer and wine.

There is a growing group of homebrewers in Alaska I hope to learn from. There have been many more organic and Alaskan local products and flowers added to our local grocery store. My dream is to add a greenhouse to our home in the next two years. Today, 85% of our total food cost is now absorbed by hunting, fishing and gathering. I'm teaching my three children how to hunt, fish and gather and integrating them into the farmer, fisher, hunter folk cultures. One day they will take care of themselves, each other and hopefully their dear old parents. A large part of this education includes the polysemic knowledge, embodied change and positionality and understanding of environmental condition that is present in all these things. We let nature create a journey and I've found hope in the sky by hunting for things that I would have never expected to come up while I'm engaged in the harvest. For example, I now hunt the aurora. I'm part of a group of people who chase the northern lights. If there was ever something that gave me hope it would be the light that dispels the long dark nights of winter.

I said at the beginning of this chapter that I'd describe a heuristic. What is a heuristic? A heuristic allows the learner to learn with experience. It is not perfect and it doesn't need to be.

It is designed to solve problems, and in the case of the oppressive food paradigm and the culture of poverty implicit with consumerism, it can be a source of guidance to navigate the status quo when more rational guidance fails. In other words, in the absence of reason and in absurdity use a heuristic to make your decisions about finding meaning in the act of conformity and suspending reason in the face of your harvest. I hope this heuristic enlivens your thinking.

Food Harvest Heuristic

I dream.
I embrace risk.
I embrace adventure.

Catching bugs thrills me.
I hear bugs in the summer.
I notice the differences in bugs.
I write down insect differences.
I take photographs of plants.
I engage in citizen science.
I keep personal notes.
I read nature books.
I collect seeds.
I harvest.

I participate in the celebration of food.
I have relationships with those who grow food.
I feel a sense of belonging in the food community.
There is a nostalgia in fruit, vegetables, and flowers.
I seek out the inquisitive places where there is curiosity.

Social media provides a glimpse into others' dreams.
I read farming, hunting, fishing, or gathering books.
Nature is special and sway hopes for the future.
I know the geography, geology and waterway.
I learn from my pain, stings, and ailments.

I know the lay of the land.
I talk with people who fish.
I talk with people who hunt.
I live for others through food.
I know the beginnings of food.
I know what happens to waste.
I enjoy holding a fishing pole.
I talk with people who grow.
I talk with people who farm.
Recreation involves food.

I chase the wind.
I take different paths.
It feels good to harvest.
I love to observe animals.
I learn about hunting ethics.
I seek out other traditions
I meditate to find peace.
I sharpen my practice.
I remain thankful.

I stand in water.
I feel the cool.
I feel despair.
I compose.
I engage.
I hope.

References

Freire, P. (1970). *Pedagogy of the oppressed*. New York: Bloomsbury.

MacDonald, K. (2007). Cross-cultural comparison of learning in human hunting: Implications for life history evolution. *Human Nature, 18*, 386–402. https://doi.org/10.1007/s12110-007-9019-8.

Mueller, M. P. (2015). Alaskan Salmon and Gen R: Hunting, fishing to cultivate ecological mindfulness. *Cultural Studies of Science Education, 10*, 109–119. https://doi.org/10.1007/s11422-014-9645-5.

Mueller, M. P. (2014). A theory of socioEcological characteristics for food mindfulness. *Brazilian Journal of Research in Science Education, 14*, 315–329.

Mueller, M. P. (2009). Educational reflections on the "ecological crisis": EcoJustice, environmentalism, and sustainability. *Science & Education, 18*(8), 1031–1055. https://doi.org/10.1007/s11191-008-9179-x.

Singer, P., & Mason, J. (2007). *The ethics of what we eat: Why our food choices matter*. New York: Rodale.

Michael Mueller is a professor of secondary education with expertise in environmental and science education in the College of Education at the University of Alaska Anchorage. His philosophy focuses on how privileged cultural thinking frames our relationships with others, including nonhuman species and physical environments. He works with teachers to understand the significance of cultural diversity, biodiversity, and nature's harmony.

He is the co-Editor-in-Chief of *Cultural Studies of Science Education*.

Index

© Springer Nature Switzerland AG 2020   199
J. B. Pontius et al. (eds.), *Place-based Learning for the Plate*, Environmental
Discourses in Science Education 6,
https://doi.org/10.1007/978-3-030-42814-3

Printed by Printforce, the Netherlands